THE HANDBOOK
OF EARTH SHELTER
DESIGN

THE HANDBOOK OF EARTH SHELTER DESIGN

BY MIKE EDELHART

Drawings by Efren Rosado

A DOLPHIN BOOK, Doubleday & Company, Inc.
Garden City, New York   1982

Photo and Illustration credits
The illustrations in this book were drawn by Efren Rosado from plans and ideas provided by many architects. Thanks to them all.
The photos on pages 154, 158 and 165 are by Rainbow.
The photo of the SunEarth House on page 147 is reprinted with permission of Rodale's *New Shelter* Magazine. Copyright © 1980, Rodale Press, Rodale Press, Inc. Photo by Carl Doney.
The photos of the Webb House on pages 140 and 141 are reprinted with permission of *Organic Gardening* Magazine. Copyright © 1978, Rodale Press, Rodale Press, Inc.
The photo of the Pendell House on page 152 is used courtesy of *Earth Shelter Digest & Energy Report.*
All other photos were provided for publication by the architects whose work they picture. Grateful acknowledgment is made to them all.

BOOK DESIGN BY SYLVIA DEMONTE-BAYARD

Library of Congress Cataloging in Publication Data
Edelhart, Mike.
    The handbook of earth shelter design.

    (A Dolphin book)
    Bibliography: p. 192
    Includes index.
    1. Earth sheltered houses—Design and construction.
I. Title.
TH4819.E27E42   728

AACR2
ISBN 0-385-17308-3
Library of Congress Catalog Card Number 80–2956

For Kathy Keeton, Ben Bova and Ande Dorman, who brought me to this fascinating subject and then brought me to New York.

This book would have been impossible without the splendid help of the following people:

Diana Bingham for working wonders with a manuscript that looked like a road map with hiccups.

Malcolm Wells for his hospitality, wisdom and heaps of information.

Eric Rosen for making preliminary contacts with architects while I was struggling along elsewhere.

And my editor, Barry Lippman, for shaping the book out of my raw enthusiasm and watching over it with admirable calm under difficult circumstances.

A special thanks to my wife, Penelope, who put up with my preoccupation and periodic disappearances while I put the book together.

All work passes out of the hands of the architect into the hands of nature to be perfected.

HENRY DAVID THOREAU

CONTENTS

INTRODUCTION

This is a book about a new way of building, living and working. It is aimed at the people most likely to spearhead the wider development of an important new idea, especially at homeowners and business people who aren't afraid to think in new and dramatic ways—especially when it can save them big money.

The idea that spurred the book was this: Now is a superb time for Americans to consider living with the earth. Not under the earth or in the earth but with it.

There is no good reason today why houses, offices and earth must be thought of as separate entities. It may be traditional to say that our houses should sit on top of the earth, but it isn't an eternal verity and, in fact, it doesn't make much sense in today's energy-hungry world.

If buildings and earth are separated, then we are, in effect, building in the air. And air is a thief responsible for the loss of heat and moisture from homes. It erodes their surfaces and loosens their joints. It makes them inefficient and shortens their lives.

Earth, on the other hand, serves as a warm blanket for a building. It stores heat and buffers the walls from cold, rain and wind. It all but wipes out the wear and tear that destroy air-exposed structures. A building designed in harmony with the earth will be far longer-lasting and inexpensive to operate than any surface building of a similar size.

But doesn't building with earth blankets mean that we'll be moving back into the caves or, even worse, down into dingy bomb shelters and subways tunnels? Hardly. Pioneer architects in the field have developed techniques that allow buildings covered with earth to seem as airy and open as any wood frame colonial. Our growing sophistication in designing with the sun in mind has greatly increased the potential for keeping buildings sunny and bright even with limited window space. And a house doesn't necessarily have to go down into the earth; the earth can be molded up around the house in myriad ways to strike a balance between the owner's desires for appearance, on the one hand, and ambience and energy savings on the other.

The name for the movement that has sprung up around this idea is earth sheltering. It is a good name because it conveys the essence of what the architecture is about. The earth doesn't bury, protect or smother. It shelters. And it does so in a gentle, flexible fashion in keeping with the needs of most modern American life-styles. Earth shelter isn't a radical idea; it's a comfortable one.

Also, earth shelter is most definitely not a specific school or style of architecture. At best it defines a loose group of styles developed by many different architects for many different reasons. Their buildings don't look alike or work alike. They use different terrains, different materials and different concepts. There is no such thing as the typical earth-sheltered building, just as there is no such thing as the typical air-exposed building. The realm of earth shelter certainly offers as wide a range of design possibilities as the uncovered world—probably more.

But the core of the movement isn't the shape of the buildings as much as the attitude behind them. In earth shelter the ground around us is seen not as a problem but as a blessing. The goal is not to make the land conform to the building but to design the building to get the most from the land. The architect's hope isn't to create a structure that stands out for miles and catches the eye instantly but one that blends so effortlessly into its surroundings that you could almost literally fall over it before you knew it was there.

It is fashionable to decry sterile, concrete suburbia. Well, with earth-sheltered ideas such a monstrosity would be impossible. What you would see in an earth-sheltered subdivision would be trees, grass, hills and meadows. And if you looked very, very closely you might make out some unobtrusive windows and solar panels and doors amid the greenery. Some of the hills would turn out to be walls. And some of the little valleys would, on close inspection, become patios and atriums.

Earth shelter is a more natural way of thinking about putting up buildings. In fact, it makes the phrase "putting up" obsolete. A better word now might be "formed." Earth shelter buildings can't just be put up; they must be formed. They can be formed in many different ways. Some are built into hillsides. Others have earth piled onto their roofs or built up along the sides. Large sub-surface atriums are the heart of one style, while virtually no surface contact at all is the key element of another. The interiors of most, whatever their exterior design, look remarkably like the best surface buildings around today. They do not, by any stretch of the imagination, look dank, dark or dreary. Many of them look downright uplifting because light has been so craftily and carefully channeled throughout.

Achieving the exciting effects of earth shelter is no easy matter. A building that can't just be thrown up can't be thrown together either. Construction and design of an earth shelter take special care and special knowledge. The benefits of building with the earth come hand in hand with an extended family of problems and considerations that must be handled wisely.

This book reviews all of those important topics. It provides a solid base of information for anyone interested in living in, or just learning about, these new-concept buildings. It won't, on the other hand, turn anyone into an instant expert on building an earth shelter. Building with earth is a process of great technical finesse and exactitude. No book for general readers could offer the details and mathematical calculations required to do it right. That is the province of technical manuals and architectural journals.

No one who has any trouble understanding these complex publications should even consider attempting to design or build his or her own earth shelter. And certainly no one should attempt it with this book alone.

The Handbook of Earth Shelter Design gives you the big picture of the movement, its problems and potential. It introduces you to the leading figures in this new way of architectural thinking and takes you inside many of the movement's finest buildings—and a few of its simplest, plainest ones, too. It surveys the extensive but not widely heralded growth of earth shelter commercial development and offers an unmatched list of references, in the form of appendices, for those who want to know still more.

Earth shelter is not a fad. It is not an architectural game or a dilettante's plaything. It is a practical, workable idea developed to mesh perfectly with these energy-parched, land-scarce, nature-deprived times. More than any other architectural idea alive today, earth shelter makes sense.

THE HANDBOOK
OF EARTH SHELTER
DESIGN

PART ONE

WHAT IS EARTH SHELTERING?

In Price and Sylvia Lierly's yard in Sapulpa, Oklahoma, it will always be spring. It will never rain. It will never be windy or cold. Even when it is snowing outside, the Lierly's yard will be warm and bright because even though it is outside their house, it is not outside.

It's about ten feet underground. Covered with dirt. Encased in concrete. Buried, along with the house, in a huge, man-made cavern. And yet trees grow there. Flowers bloom. Breezes blow. The air is fresh and suffused with sunshine.

The Lierly's house was designed by a tough-talking Texan, Jay Swayze, who has created, over the past few years, some eighteen of these underground dreamscapes. Swayze is one of a growing number of architects and homeowners who believe that habitations belong back in the earth.

The Cape Cod home of architect Malcolm Wells is also sheltered by the earth, but it is entirely above ground. Wide windows cover three sides of the long, airy ranch-style home. The blue expanse of the Atlantic Ocean rolls away from the base of the cliff on which it stands. Light and breeze are everywhere in the house, which is appreciably cooler on a hot summer afternoon than the woods outside.

The earth in Wells's house is on the roof. Three tons of it hold the summer's heat to buffer the house in winter and the winter's chill to cool the house in summer. The walls are buttressed by slopes of earth that rise outside to just below the windowsills. An enormous amount of insulation lies behind the deceptively delicate-looking walls. The house is a sophisticated temperature-modulation chamber, carefully crafted for energy efficiency all year round. But a visitor doesn't feel like he or she is in an insulated cave at Malcolm Wells's house; all the guest feels is the beauty and openness of a splendid design.

Whether they go truly underground, as Swayze would have it, or merely mold their homes into the surface with an earth-tempered design, as Wells prefers, Americans are unquestionably wrapping themselves in earthen blankets—and loving it.

There are at least three thousand underground homes in the United States right now—and probably quite a few more. They may be found from Cape Cod to the California coast and range from homemade Hobbit holes to multistoried mansions. One fellow is even selling cave-home construction franchises.

Individual homes are not solely involved in this burgeoning movement. Whole developments are going underground. Hewn 150 feet into Iron Mountain, in the upper Hudson River valley, for instance, is Safe City, a subterranean complex that includes warehouses, archive vaults, and two-story office blocks. This development was built with security in mind, being located inside an old cement mine, and includes luxury apartments that wealthy tenants can occupy in the event of a nuclear attack.

Both more practical and more impressive is the vast commercial catacomb beneath Kansas City, Missouri: twelve square miles of industrial park on two levels between fifteen and thirty yards underground. This subsurface warren used to be a limestone quarry that was mined horizontally into the rock face, leaving behind huge "rooms."

Today more than two dozen companies use this space for everything from plush offices to factories, bringing some two thousand workers underground each day. Tenants include the Inland Cold Storage Company, which can house in its refrigerated caves one pound of frozen food for each American, and Brunson Instrument Company, manufacturer of the surveying equipment used on Apollo missions, which found the vibration-free underground atmosphere ideal for producing precision optical instruments.

Down the road, in Omaha, Nebraska, one of America's more staid companies, Mutual of Omaha, is going subsurface. Mutual's new office will be an ambitious three-story underground structure with a massive glass dome at the surface. Mutual is going down because going underground allows the company to use land that would otherwise become a parking lot. The twenty-thousand-square-yard subsurface project economizes on space, and the street-level dome relieves the dowdy appearance of downtown Omaha. It's a development that seems to have made everyone involved happy.

Underground residential subdivisions are spreading their roots around the country, too. Projects are currently in the works in Texas, Michigan, Missouri, and Ohio, and more are on the way. The ultimate lure of subsurface subdivisions will be exurban greenery combined with suburban conveniences. Designers of one project near Dallas, for instance, plan to allot 86% of the development to greenery, compared with the 50% typical of surface residential subdivisions.

ENVIRONMENTAL ADVANTAGES

The main reasons among architects for turning to the earth are environmental. Malcolm Wells, who serves as a sort of preaching Saint Paul for the movement, has perhaps best expressed it. According to him, underground architecture

- offers silent and vibration-free living, along with absolute privacy;

Architect David Wright's house is snuggled into a hillside overlooking California's Pacific Coast.

- presents natural substances instead of roofing materials, to the sun, with plants that give off oxygen and that even provide food, instead of the useless heat of typical roofs;

- produces greenery and wildlife habitats in place of asphalt and urban blight, in keeping with the concept of planet earth as the ultimate national park;

- eliminates extremes of weather and temperature;

- does away with (once rooftop greenery is in place) the waste of energy and time required for yard maintenance;

- saves the precious rainwater that surface homes waste, allowing it to percolate and feed the earth instead of running off and causing erosion;

- allows the recycling and subsequent utilization of household organic wastes to increase the yield of the home's biosphere;

- permits houses to be built closer together than they are in most modern suburbs, with less feeling of crowding and less impact on the land.

Kenneth Labs, another earth shelter architect, lists a few more prosaic advantages:

- protection from storms and other climatic extremes;

Malcolm Wells's fanciful "A Random House" combines sophisticated construction and design with a totally natural appearance.

- stability in the face of natural disasters, such as tornadoes and fires;

- ease in separating unrelated systems, such as traffic (surface) and pedestrians (subsurface);

- reduction of visual pollution and the sensory overload a compact community of unsheltered buildings brings with it;

- maintenance benefits and lower insurance rates, because the house is sealed from weather and is made of noncombustible materials.

ENERGY SAVINGS

Among homeowners, however, the biggest advantage of underground housing is the extraordinary energy savings. "We pay one third of what we would have paid in a regular structure," says underground-home owner Pat Clark, of frosty River Falls, Wisconsin.

Mutual of Omaha figures the heating cost for its subsurface offices will be one fourth of what it would cost above ground. Most of the offices' heat will come from body heat, lights and kitchen waste steam.

Extensive studies by Dr. Thomas Bligh of MIT, a founder of the American Underground Space Association when he was at the University of Minnesota, predicted a typical energy saving of 75% with an earth-covered house. Bligh states, "In no way can improved insulation on an aboveground building begin to compete with subsurface structures from an energy-conservation standpoint."

The reason underground housing saves so much energy is not because earth is a good insulator. In fact, as Wells notes, "earth is a lousy insulator. Urethane foam is perhaps ten to twenty times as good. But," Wells points out, "earth . . . is a great moderator of temperature change. Warm it up, and it stays warm a long time."

Earth does not react as fast, or as severely, to temperature change as air does. This means that a house buried in earth has a much narrower band of temperature variation to cope with than a surface structure has.

If, for instance, air temperature on the surface ranges from 0° to 95° F, four yards down the temperature of the earth will vary only from 50° to 65° F. The earth serves as a warmer in winter and as a cooler in summer, tremendously reducing the load on a home heating system.

There is an additional benefit called the thermal flywheel effect. Because the earth loses heat slowly, the coldest earth temperature will lag several months behind the coldest air temperature. This means that the earth around an underground house is still warm and slowly cooling in January and February. It hits temperature bottom only in April, by which time the need to meet the peak heating demand is past. The same goes for summer cooling: The earth doesn't hit peak temperature until the summer's heat is already waning.

Other enticing features of earth-covered homes include their incredible durability and stability. Jay Swayze notes that "it's nature that destroys shelter, even with continuous maintenance. By putting dwellings underground, we are attempting to put shelter in harmony with nature. Protected from the ravages of weather, an underground house can last forever." Down beneath the surface, houses avoid the onslaughts of tornadoes, hurricanes, and other surface destroyers. They even stand up better to earthquakes than aboveground buildings, although they're surrounded by tons of dirt.

John Barnard, of Marstons Mills, Massachusetts, one of the most influential underground architects, explains: "Underground houses fare far better in earthquakes because the biggest damage to surface buildings is caused by a whipping action. The earthquake twists the building and then whips it back, tearing it apart. Underground buildings can move only as far as the earth does. They can't whip."

The maintenance feature strikes a primary chord with older buyers. "A lot of people are coming to us, thinking to retire. They want a place that has minimum upkeep and maintenance," says John Hand of Moreland and Associates, a Texas earth shelter developer. He notes that surface homeowners can count on

Hillside House, designed by William Morgan, is literally surrounded by the crest of the hill. This saves energy, wear and tear, and presents the eye with a view of a hill, not a building.

needing high-cost repairs just when they are ready to retire on a decreased income. Underground homes suffer far less deterioration. The glass, concrete and anodized aluminum they are made of last virtually forever. And they are shielded from the erosive elements.

Weather protection does not merely increase home longevity, it makes homes safer. One stormy afternoon a tornado was spotted just a few miles away from Frank Moreland's earth-sheltered model. "Twenty-two neighbors decided it was a great time to visit."

For some, earth shelter represents a way to preserve scarce land for purposes that make life richer and healthier, such as gardening or recreation. They note that the homes are just as ambient as surface dwellings but limit land use options far less.

DISADVANTAGES

Earth-sheltered houses, though, do have their drawbacks. No architectural form is entirely free from troubles. Among the potential problems and shortcomings of earth-sheltered homes (which will be discussed in more detail later) are:

- restrictions imposed by climate, region, and local geology;
- difficulties with condensation and high humidity;
- lack of visual identity, image and "presence";

- less direct and perceptible modes of access;

- linkage difficulties between surface and underground facilities;

- potentially higher initial construction costs;

- dubious palatability and public acceptance in some communities;

- difficulties for future home expansion;

- long waits for most economic gains;

- possible psychological drawbacks.

Despite kneejerk fears of dark, closed-in spaces, people who move into earth-sheltered houses don't suffer from claustrophobia, cabin fever, or any of the negative effects many people imagine. Rather, "they just feel it's a natural way to live," says underground-housing expert Gordon Moore, of the University of Missouri at Columbia.

"That's true," agrees Barbara Webb, who, with her husband, Larry, has lived in an underground house near Drexel, Missouri, for about two years. "You don't feel closed in. You don't feel uncomfortable. It just feels right. Like a house where people live."

In this interior passive solar design requirements have been handsomely integrated with the surroundings. The tall column (left) captures air warmed in the greenhouse (right) and circulates it through the house.

An earth-sheltered home doesn't mean a dingy, dark interior.

An earth-sheltered house can have a conventional appearance.

The interior of an earth shelter can be multilevel and as open as the imagination of the designer.

An earth-sheltered library/study. Just because the windows are small doesn't mean a room must feel closed in.

The bold corners and unusual structural features of an earth shelter can be muted by warm, traditional decorating ideas and furnishings.

This earth shelter kitchen could be in any fine home. It lies at one end of the more open south-facing wall so windows can let in plenty of morning sunshine.

The view from this family room is toward the greenhouse, which covers the south-facing wall. The greenhouse provides privacy and a grand view. It also collects solar heat.

AN EARTH-SHELTERED HISTORY

The current interest in underground living as an American alternative may be new, but underground living itself is as old as civilization. The textbooks say humankind left the caves for good in search of new food supplies millions of years ago. But in reality vast populations have lived beneath the earth in all eras and regions. As Wells has said, "I can just picture some smelly old brute, club in hand . . . surprised at seeing a cave for the first time . . . It must not have been too many minutes later that underground architecture was born. Despite great advances in the techniques of building above ground, man has never completely abandoned underground construction. Remains from every age on every continent prove that man has continued to avail himself of this most ancient of architectures."

Archaeologists have unearthed amazing subsurface settlements from the early days of civilization. Five thousand years ago, for example, human denizens of the Negev built homes underground to avoid the sun. In Tunisia the ancient Romans put street-level courtyards atop their cool, subterranean living areas.

Some American Indian tribes lived in hollowed-out cliff faces or in earth-covered lodges. Indians of America's Southwest constructed subsurface prayer chambers called kivas. Their earth-heat dampers, along with savvy natural ventilation, kept these religious hideouts cool and comfortable during scorching desert summers.

In Tunisia, thousands have lived for centuries in warrenlike towns nestled in the shadows of the hot lowlands. Homes in villages like Matmata are cut into the earth off of belowground courtyards, sometimes as much as thirty feet beneath grade. The natives spend virtually their whole lives in these sunken yards and the tunnels that connect them, away from the heat and fierce winds at the surface.

China has had an earth-sheltered population in the millions for generations. Living goes on snugly under a blanket of mountain soil. The weather in China's mountains can be fierce, but the earth-sheltered housing pattern protects inhabitants from climatic extremes. Some units in China are even situated to take advantage of the low winter sun—a precursor of today's passive solar designs. One anthropologist reported that "not only habitations but factories, schools, hotels and government offices are built underground."

Of course, in polar regions expeditions burrow beneath the ice to escape frigid winds and to conserve heat. The principle involved is much the same as in earth sheltering.

Among Caucasian Americans, however, earth-sheltered housing never caught on because settlers used the new continent's seemingly inexhaustible supply of timber to build houses on the surface. Abundant amounts of land and resources effectively dampened prosperous America's interest in subsurface dwelling until the land-and-energy-tight 1970s.

Except, that is, for one Italian immigrant who bucked the trend in spectacular fashion beginning in 1908. Baldasare Forestiere spent thirty-eight years carving an incredible seven-acre maze of sixty-five rooms, courtyards, grottos, and gar-

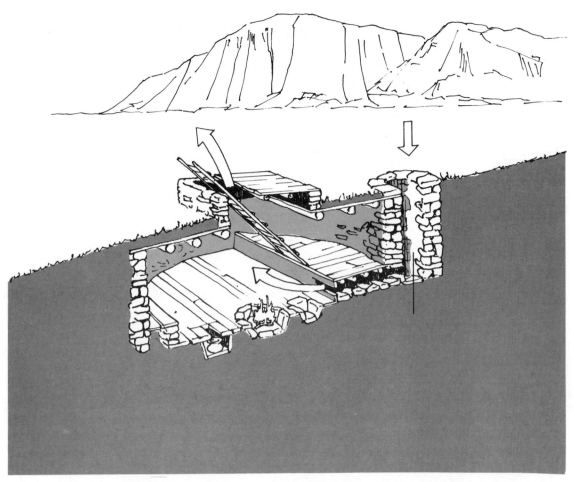

American Indians knew about the heat moderation of earth shelter. Their kivas, or prayer halls, were dug down to keep them cool during desert summers.

dens ten to twenty feet beneath Fresno, California. Light enters this subsurface wonderland through atriums and rooms strategically placed to receive the morning or afternoon sun. The sunken courtyards are rife with plant life, including one tree twenty feet underground that bears seven kinds of citrus fruit on grafted branches.

No one is certain about why Forestiere went to all the trouble. But his life-work—he was still working on it when he died in 1946—remains an uncanny presentiment of the future.

The one form of earth shelter dabbled in by our ancestors was sod. This, however, was a building material that denoted extreme privation. You'd only build a sod house if nothing else was available. And sod offered lousy protection from the elements. It let in the cold wind and held heat like a sieve. The extreme discomfort of sod homes was bemoaned in a verse from the Kansas tune "The Little Old Sod Shanty":

The hinges are of leather and the windows have no glass,
While the board roof it lets the howling blizzard in,
And I hear the hungry coyote as he slinks up through the grass,
Round my little old sod shanty on my claim.

Our bad experience with sod may have scared Americans away from other earth-sheltered possibilities for generations.

Throughout history the commonest use of land-integrated buildings has been for defense. Petra, in present-day Jordan, protected itself from marauders in the days before Christ by hewing to the cliffs. "In all parts of Petra," writes Philip C. Hammond, "every available rock has been worked to a vertical face, sometimes left plain, but more often sculpted as the facade of a temple, a shrine or a dwelling. Behind each facade is a large chamber hewn in the rock and entered through a tall rectangular doorway."

In modern times earth shelter brings to mind the atomic bomb survival chambers of the 1950s—a negative image that still plagues the movement.

American military forces and those of other developed nations have built extensively underground. The North American Air Defense Command operates from a city of eleven buildings, some three stories high, built beneath Cheyenne Mountain in Colorado. The buildings are spring-mounted to ensure that they'll withstand the vibrations of an attack. The American Wartime Communications and Command Center is buried beneath Battle Mountain, near Washington, D.C.

Perhaps this long genealogy of underground living helps explain why modern Americans who are trying this alternative find it surprisingly natural and comfortable. Somewhere in our genes lurks an ancestor who lived surrounded by earth and who is happy to see us return.

MODERN EARTH SHELTERS

Of course, few people today would choose to live in our ancestors' closed, unadorned caves. So earth shelter designers have concocted several plans that provide many of the benefits of total submersion with free access to the surface.

The most popular of these is the atrium house, first designed and championed by John Barnard. "I'd been fascinated with underground housing since I was a boy," Barnard, who is sixty, explains. "My dad was an architect, and he would always say that if a house were deep enough in the earth, it would be 60° all year round. So why not build houses down there?

"I mulled it over for a long time. My wife's response to trying an underground house was, 'I hope you and the mole you marry after I leave you will both be very happy.'

The ancient city of Petra protected itself against enemy attacks by hiding within a gigantic natural cavern. Security remains a primary asset of earth shelter today.

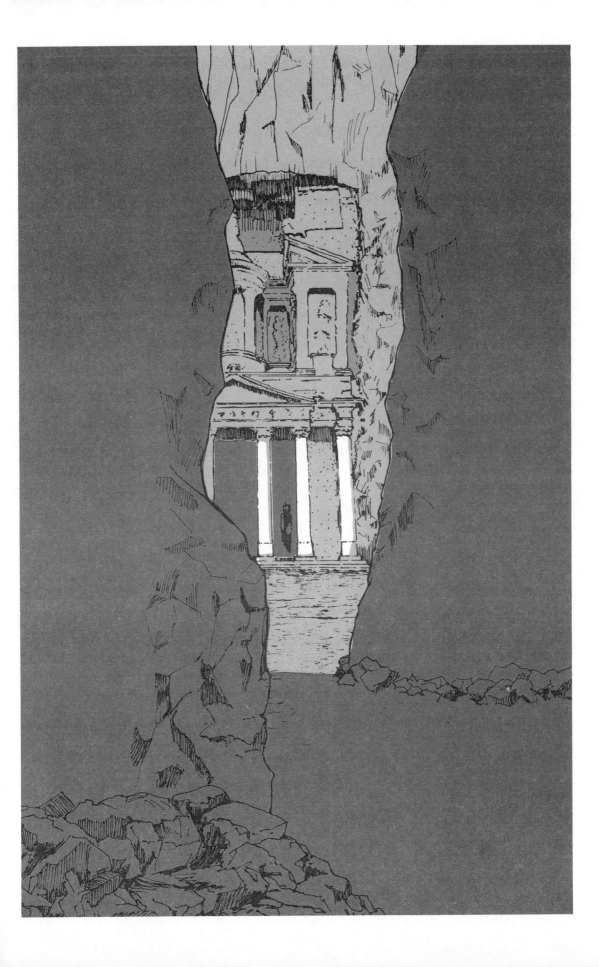

John Barnard's atrium-sheltered Ecology House has become the prototype for the most popular earth shelter style.

"Then, on a trip to Pompeii, we were sitting in a Romanesque-style cafe with an atrium. And it dawned on me that if it were all dropped ten feet underground, you'd have the best of everything. My wife bought the idea, so I went ahead and built the house."

The result was Ecology House, completed in 1973 as Barnard's vacation home. It features a living area that surrounds a sunken atrium. Light reaches every room in the house through walls facing the open courtyard. Every room has a view of the sky and the outdoors, yet the house as a whole retains most of the energy savings, privacy, and silence of completely buried designs.

Initially Barnard realized a 25% saving on construction costs over a similar-sized aboveground house. Since then he has paid only one fourth as much for heat.

The atrium-style earth shelter house has gone on to become the most popular kind of underground design. Barnard is justifiably pleased. "Everybody in town thought I was crazy when I built it," he says slyly, "but electricity was 1.9 cents a kilowatt then. They don't think I'm crazy anymore."

Wells thinks the atrium design has major drawbacks. For example, he feels that the layout means "you just end up looking across at someone else's window." Wells prefers houses that are built into hillsides.

The hillside house is more cavelike than the atrium home, being recessed into a south-facing hillside. Only the front wall has any contact with the surface. The southern exposure orients the front wall, which usually contains a lot of window space, to strong sunlight, which fills the upper level of the house. The sunlight can also drive solar-heating units to warm the building.

Since earth flows around the hillside house, it literally becomes part of the landscape. In that sense the house becomes a living entity. As Wells once wrote of his old underground office near Philadelphia: "I don't know what color my office roof will be this fall. Last fall it was solid yellow, done in masses of wild sunflowers. It was so intense it seemed to pulsate. Then, a few weeks later, it turned a hundred shades of gold as frost after frost went to work on it."

Wells also argues that hillside homes create the best housing on the worst housing sites. "Eroded farmland or strip mine sites would be good," he states. "Valuable, rich flatland could be left free for parks, farms, and other uses, rather than for highways and apartment complexes."

The most unabashedly underground of the new designs are those of Jay Swayze, who works out of the small town of Plainview, Texas. Swayze describes his homes as dreamworlds. "A man's home is his castle, and we make it his dream."

To a large extent Swayze lives up to his claim. He builds posh suburban estates inside gigantic underground caverns replete with space-age accoutrements. These create an ever-changing Elysian environment that includes both an "inside" and an "outside" beneath the surface.

A hillside home is covered with a blanket of earth that molds it into the contours of the surrounding countryside.

Swayze houses feature his patented air system, which provides a constant breeze around the house. Sophisticated skylights and fiber optics create an impression of sunshine and the feeling of day passing, much as a surface dwelling does. Sensitive environmental controls make it possible to grow plants, even trees, in the "front yard" of the Swayze design.

"In one of our houses," Swayze states, "you get the best of both worlds. You get the security and privacy and energy savings of living underground. And you get an improved surface world. On earth you have seasons. But in the earth there's no reason why you can't have spring all year."

While Swayze emphasizes control, other earth shelter architects stress designs that make minimum impact on the environment. "The essence of the ecological argument for underground space," writes Kenneth Labs, "is that its use can minimize a building's impact on the local biotic community and natural processes. By building beneath the surface, or by utilizing soil and plant cover as an integral part of a building's insulation and structure, one provides the opportunity to re-establish a plant community and its associated wildlife habitats. These, then, provide for the retention of beneficial biological controls, greater species diversity, and reinforcement of the pre-existing integrity of the local ecosystem. The earth-building practice also allows nature to process rainwater in its normal, unhurried way, in addition allowing man to capitalize on a host of useful functions provided by plants, for example, shading, evaporative cooling, and dust filtration."

"Underground buildings," says Malcolm Wells, "can help today's land- and resource-strapped world. They offer so much silence and privacy they can provide high-density accommodations with no loss of their inherent greenness. I see no reason at all why high-density, solar-heated, low-income housing can't be as green and as beautiful as parkland—right in the middle of the city—with far lower energy and maintenance bills than now seem possible. There's still a lot to learn about this ancient way of building, but even at this stage all the signs say 'do it.' "

James Scalise, an Arizona earth shelter expert, sums up the new-styles role in architecture today: "In conjunction with other directions of energy and environmental conservation, such as solar heating, dry waste disposal systems, natural ventilating techniques, et cetera, earth shelter certainly represents a rational and exciting approach to solving the myriad building problems faced today in the light of our dwindling resources. We may even be surprised at the quality of integration resulting from such unique approaches to creating and living in harmony with our environment. It is also possible that we might even discard our tendency to create 'pedestal architecture' in favor of being concerned about designing the 'place' and not just the building."

Andy Davis's cave home in Illinois points up one of the goals of earth shelter: to make the home blend as much as possible with its surroundings.

PART TWO

WHAT YOU NEED TO KNOW

SOIL

Below-grade buildings are cradled by soil. They must be carefully conceived to fit in with the specific particulate environment in which they lay. The structure and design of an earth-sheltered house are intricately entwined with the weight, moisture, texture and density of the surrounding soil.

First of all, soil can mean any kind of material, from tough, compact demi-rocks and clay to sand. It can be composed of an infinite variety of rock material and organic leftovers. It has liquidlike properties; both clay and sand, for example, flow. In addition, soils vary with differing environments; clay that is saturated with water is far different from bone-dry clay.

In short, soil is a complex, changeable, vital element of any earth-sheltered design, one that should be examined carefully in conjunction with a professional engineer. A below-grade house that isn't in sync with its couching dirt can literally slide out from under its owner or be squashed with impunity.

Soils are classified according to four basic types: gravel, sand, silt and clay. Generally, soils in nature are mixtures of these four components. A few such combinations are common enough to have names of their own. A mix of sand, silt and clay is called loam. Fine-grained, sandless clay is called gumbo because it has the dark, greasy consistency of that Cajun concoction. Humus is the name for the rich, dark topsoil we associate with gardens. Hardpan is earth that has been squeezed hard, almost to the point of rockiness; it will not soften when wet. Porous but stable deposits of tiny particles, like compressed chalk, are called loess. And mud, as any child's toes will attest, is an oozy mix of silt, clay and water.

The relationships among the soil's components are generally displayed in the Feret triangle. No mixture appearing in the triangle makes earth sheltering impossible, but the costs and problems increase as you near the corners. The soils at the angles are more likely to crumble, slide and break.

As you dig deeper, you find that soil isn't consistent. It follows a layered pattern called a soil profile. When an engineer takes a soil boring, he is examining this profile. The depth of various types of soil, obviously, is extremely important when planning an earth-sheltered building. The composition of the soil at the building's resting depth may not match what is seen on the surface.

SOIL CHARACTERISTICS

Different types of soil display different basic characteristics. These properties directly affect all earth-sheltered buildings and should be familiar to anyone thinking about building one.

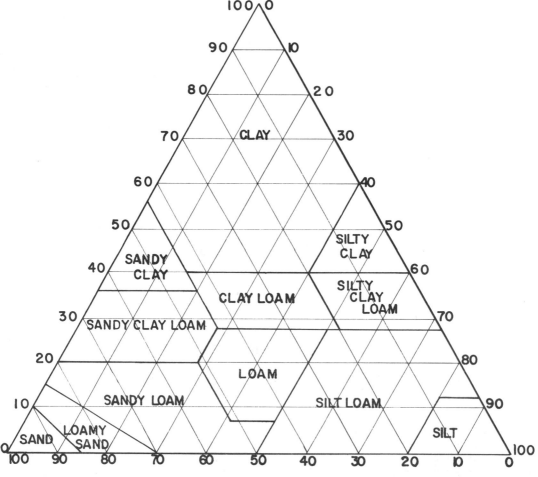

The Feret triangle relates the various types of soil. The soils near the heart of the triangle are most stable, while those along the edges tend to be quirkier.

Cohesiveness. This refers to the tendency of soil particles to hang together. When the bond between the particles is intermolecular, the cohesiveness is known as real; when moisture holds the soil together, the cohesiveness is called apparent.

Noncohesiveness. As you might expect, this refers to the tendency of soil particles to go their own way. Dry sand is noncohesive.

Plasticity. Some soils can stretch, mold and flow without ripping. All soils have some degree of plasticity; the greater the degree, the more the stretching ability. Plasticity indexes are used by engineers to determine precise amounts of relative stretchiness in different soils. The stretchiness of soil is neither good nor bad—as long as the building is designed for the conditions at hand. A poorly designed house could shift when the soil stretched.

State. H_2O has three states: solid, water and gas, depending on temperature. Soils have four states, depending upon moisture content: solid, semisolid, plastic and liquid. As the water level in soil increases, it moves from one state to the other. The points where it passes from one state to another are called limits: shrinkage limit, plastic limit and liquid limit. The pattern for states and limits is like a particular soil pattern's signature. This signature expresses how the soil will behave in all circumstances, from drought to inundation. It allows an architect or engineer to design an earth-sheltered building to handle all possible circumstances in the surrounding soil.

Angle of repose. Depending upon the internal attraction and friction of soil particles, various types of soil will naturally assume certain slopes. If you poured a bucket of sand onto a floor, it would hardly form any pile at all; but rich, moist loam would form a neat, stiff pile. The maximum angle a pile of soil will hold without slippage is its angle of repose. This controls the slope at which soil can be angled away from an earth-sheltered house and angled against walls and up from the edge of the roof. If an earth-sheltered house is put into an excavation too narrow to allow the soil to be raked back to its angle of repose, the house will be buried—whether or not the owner wants it to be. Ground covers, such as a lawn, can be used to hold soils at a steeper angle than they would attain on their own; so can some chemicals.

PRESSURE

If angles and consistency of soil were the only considerations, designing earth-sheltered buildings would be simple. Unfortunately, soil has other impacts on a sunken house. The most serious one is pressure.

When a rectangular house is covered with earth, the soil presses upon the house from all directions. The weight of the soil on the roof presses down. The slippage of the soil creates horizontal pressure against the walls. Moisture in the soil creates hydrostatic pressure that pushes up against the bottom of the building and increases the tendency of the soil to flow, increasing pressure on the sides as well.

Even areas beneath grade where soil has been cleared away suffer from soil-induced pressure. The sliding-wedge principle states that excavated earth that doesn't face horizontal pressure (as in the slope of ground angling away from a building wall) acts as a wedge pressing all of its weight down at the angle of repose. This diagonal pressure pushes against the base of the wall and the edge of the floor. It is a difficult thrust to calculate, but it is vital to take it into account when designing for in-earth conditions.

OTHER FACTORS TO CHECK

The point of this detailed review of soil properties is to stress that anyone planning an earth-sheltered project should study the soil of any potential building site carefully. It's a complex medium in which to build, with many variables that might escape the considerations of a nonprofessional designer.

The best thing you can do before starting out on any earth-sheltered project is to get your soil tested by a pro. Keep in mind that the loads against the foundation footings of a surface house are just about 2,700 pounds per foot, while for an earth-sheltered home they may run as high as 12,000 pounds. That puts a lot of strain on the soil and requires extreme care in planning the size and configuration of the all-important footings.

If the footings of an earth-sheltered house are designed inadequately for the soil conditions, the building will sink into the ground—a process called settling—and form cracks and leaks; it might even collapse. If the footings are overly large, the building will stand up just fine but will cost far more than necessary.

In the absence of specific information about a building site, architects and engineers design according to standard, quite conservative assumptions about soil and other conditions. These assumptions may be fine in your case, but they may be way off for someone else. The only way to insure that the house is designed right is to get as much information about the soil as possible.

This suggests a cautionary word about through-the-mail house plans. Many of these plans are sold with specific dimensions for footings and other structural underpinnings. You can't assume that any of these specs apply to the site you are building on. Your soil may be totally different from what the ready-made design assumed. If you follow all-purpose instructions, you could settle yourself into an all-purpose disaster.

Unless you have some engineering background, you should get professional assistance in figuring structural dimensions for the walls and placement of doors and windows. The pressures against the walls of a surface house are exceedingly light since it is surrounded by air. Even a surface house basement suffers rather light pressures; all builders and most handymen know that reinforced concrete, blocks or sturdy timber can handle the load.

But in earth shelter, at greater depths and complexities than in a typical basement, the pressures from soil weight, movement and hydrostatics will be greater

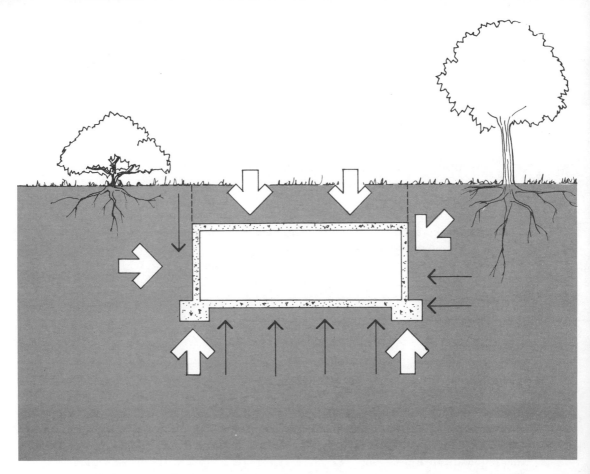

An earth shelter's soil environment presses in upon it from all directions: weight from above, slippage from the sides, and hydrostatic pressure from below.

and harder to figure. It is far cheaper to get good advice in planning than to overcompensate and pay more in labor and materials or to guess wrong and have your house mesh improperly with its surroundings.

Here are some facts you should know about soil at a site before you dig the first spadeful of dirt:

- What is the groundwater level? Where is the water table (i.e., the natural level of underground water below which you may have a drainage problem)? Do your design ideas and the soil moisture conform or clash?

- How much direct weight can the soil accept? This is called the bearing weight of the soil.

- What is the cheapest form of foundation that will insure your house's long-term viability in the surroundings?

- What are the pressures of soil in all directions and at all critical points of your design? Do they give you a margin of safety?

- How much is your house likely to settle? Will settling cause problems?

- Will the presence of your house underground cause any headaches for neighbors? For instance, if you force the water table down beneath your house, will it rise around theirs, flooding basements?

- How quickly does the soil around your house accept water and transmit it downward? This is called percolation. (For more details, see the section entitled Water.)

This list does not cover all the possible factors concerning soils, but it includes the main ones. All the pertinent facts can be obtained through a soil sampling and engineering consultation that will cost about seven hundred dollars.* The cost of not doing it could be an unhappy home.

THERMAL CONDUCTIVITY

Besides factors crucial for construction, soil also has attributes that affect the energy-saving capacity of earth shelters. A University of Minnesota study found that different types of soil transmit heat at different rates; some drag heat away from a house faster, some slower. This is called thermal conductivity. The University of Minnesota study found that the factor can change by a power of ten, depending upon how moist the soil is. Soils that are moist will pass heat more quickly than drier ones. The heat transmitted through the moisture is lost for modulating house temperature. Since the key to underground energy savings is using the earth for heat storage and slow release of heat, fast conduction of heat through the soil is a negative factor. Drainage can help make sure moisture doesn't rob heat from the soil. (For a detailed explanation see the section on Water.)

Another reason to test for thermal conductivity is expansion in the moisture that does the conducting. Highly conductive soil will tend to expand and contract a great deal. If it is packed too tightly against a wall, its expansion could conceivably crunch the wall flat. One kind of wet clay, for example, has been found to exert more than thirty thousand pounds of pressure against one square foot of vertical wall.

In winter, when all the moisture freezes, it expands even more. The swelling that results from ice expansion, called ice heaving, can crack walls and footings in extreme situations. Worst of all, ice is even more conductive than water, so it will steal more heat from the house in winter than moisture does in summer.

If soil is extremely moist, it may need a barrier of gravel or other buffer material around the walls to short-circuit the damaging thermal thievery.

* *Note:* All prices quoted throughout this book were correct at the time of publication and are subject to change.

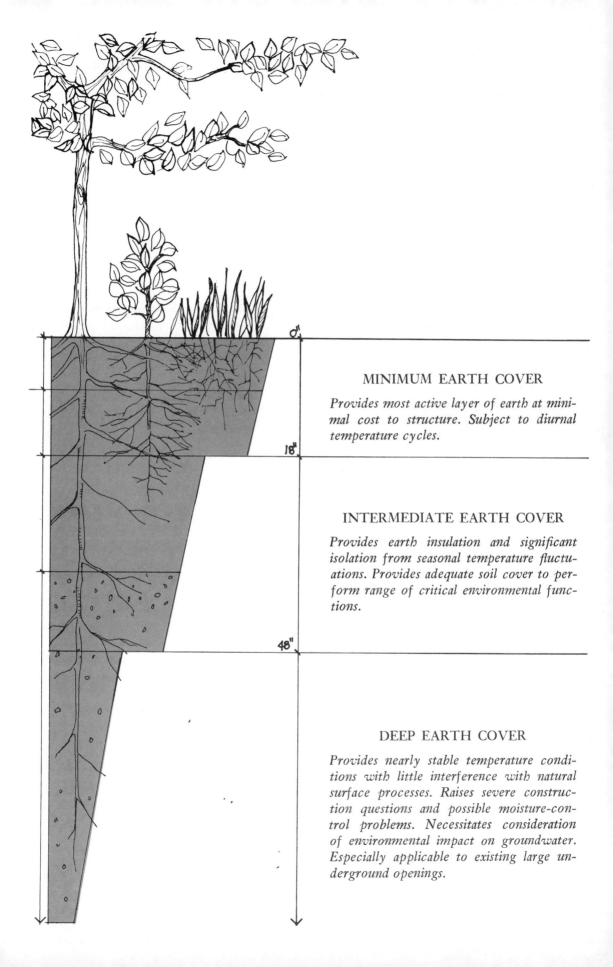

MINIMUM EARTH COVER

Provides most active layer of earth at minimal cost to structure. Subject to diurnal temperature cycles.

INTERMEDIATE EARTH COVER

Provides earth insulation and significant isolation from seasonal temperature fluctuations. Provides adequate soil cover to perform range of critical environmental functions.

DEEP EARTH COVER

Provides nearly stable temperature conditions with little interference with natural surface processes. Raises severe construction questions and possible moisture-control problems. Necessitates consideration of environmental impact on groundwater. Especially applicable to existing large underground openings.

PROBLEM SOILS

One thing that a soil survey may uncover is ledge—hard bedrock that lies uncomfortably close to the surface. Water won't drain through ledge and sewage won't percolate down into it. It has to be blasted in order to set foundations. In general, ledge represents a serious impediment to decent earth-sheltered design.

Rocky sites, as a whole, represent bad risks for earth-sheltered buildings. The problems they present can be overcome, but the cost and trouble, in the view of many architects, aren't worth it. "If someone came in with a site that was rocky, where we had to do a lot of blasting," says John Hand, "I'd tell them to choose another site, because they'd blow a lot of money trying to knock all that rock out of the way."

The other severe problem soil for earth-sheltered homes is wet clay. It's slippery, which means that it moves, and the house will tend to slide along with it. It's moist. It's heavy but doesn't support much weight. It may hold a near-vertical angle without shifting, but clay squeezes; eventually that seemingly solid vertical will come to bear on the house sides and the weight of it will squeeze clay up against the house floor.

No soil type, however, is an absolute barrier to an earth-sheltered house. Information in hand and good planning can handle any situation. A site with ledge might be suitable for an above-grade house that is bermed—with soils piled up—rather than the house being dug down. A clay site may be handled by larger footings and careful waterproofing. As John Hand says, "If we have soil tests done, if we know what kind of environment we are dealing with, we can take care of just about any site. We wouldn't build in a floodplain, but other than that, we've handled everything anyone's brought us so far."

SOIL MISCELLANY

- Soil, says Malcolm Wells, goes into shock during construction. It gets bruised and surly. You can help it generate plant life more quickly by treating it with lime, fertilizer and judicious watering as soon as construction is finished.

- Roof soil that is raked back from the edge at its natural angle of repose shouldn't erode while plants are taking hold. Hay bales held by bricks can also hold the soil in place until roots can do their thing.

- When checking over a site, make sure you look for, and ask about, old foundations or other underground construction remains. They might not bother a surface dwelling, but they could cause frustration and extra expense for an earth shelter.

- You can also learn a lot about local soil by talking to neighbors and to any construction crews that happen to be working in the area. You can get a quick notion of soil conditions by looking at excavations in the area, such as along roadsides or railbeds.

■ If you want to encourage plant life on the roof, here are five typical formulas for thriving rooftop planters:

<table>
<tr><td>**MIX A**</td><td>**MIX B**</td></tr>
<tr><td>⅓ loam</td><td>⅓ loam</td></tr>
<tr><td>⅓ peat moss</td><td>⅔ topsoil</td></tr>
<tr><td>⅓ topsoil</td><td></td></tr>
</table>

Note:
Since peat moss is highly absorptive of water, in some cases it may be desirable to increase amounts of topsoil or to add sand, as in the following. (These mixes are exclusive of nutrient additives and conditioners, such as lime, bone meal, etc.)

<table>
<tr><td>**MIX C**</td><td>**MIX D**</td></tr>
<tr><td>½ loam</td><td>¼ topsoil</td></tr>
<tr><td>¼ peat</td><td>¼ peat moss</td></tr>
<tr><td>¼ sand or</td><td>½ coarse sand</td></tr>
<tr><td> perlite</td><td></td></tr>
</table>

MIX E

2 parts hypnum or sphagnum peat
3 parts Haydite or Basalite aggregate
  (⅜-inch to ⚹8 screen)
3 parts Haydite or Basalite (⚹8 to 0)

WATER-HOLDING CAPACITY OF SOIL (AS % OF WEIGHT)

SOIL TYPE

coarse, sandy soil	15–30
light loam	22–34
stiff clay	36–50
sandy peat	53–60
peat moss	25 X
vermiculite	10–12 X

UNIT WEIGHTS OF SOIL FOR DIFFERENT CONDITIONS

SOIL AND CONDITION	WEIGHT (pcf*)	SOIL AND CONDITION	WEIGHT (pcf*)
CLAY		**LOAM**	
dry, hard	137	loose dug	75
very dense	125	in situ, dry	80
moist, loose	115	in situ, wet	120
silty, dry	100		
plastic	100	**HUMUS**	
dry, and gravel	100	dry	35
organic	88	wet	82
GRAVEL		**SUBSOIL**	
wet	120–125	in situ, dry	110
dense	100–120	in situ, wet	125
dry, loose	90–105		
		LIGHTENING AGENTS	
SAND		coke, dry	40
packed, dense	125	coke, wet	50
wet	120–125	vermiculite, dry	3.5–6
10% moisture	120	vermiculite, wet	35–75
fine, dry	100		
dry	90–105	Styrofoam	2
		Dorovon	1
LOESS	100	perlite (Perloam)	8
SILT	115	**"EARTH"** (excavated)	
		dry, packed	95
PEBBLES	110	moist, loose	78
		moist, packed	96
PEAT	70	mud, flowing	108
		mud, packed	115
CRUSHED STONE	100		
		WATER	62.4
"EARTH" (excavated)	76		
dry, loose	95		

* Pounds per cubic foot

SLOPE ANGLES OF REPOSE, COMMON GRADING PRACTICE

SOIL AND CONDITION	IN°	AS RATIO
CLAY		
dry	(30)	1.75:1
damp, plastic	(18)	3.00:1
firm	45	(1.00:1)
wet	16	(3.50:1)
SAND		
clean	(33)	1.50:1
dry	38	(1.30:1)
wet	22	(2.50:1)
GRAVEL	(37)	1.33:1
EARTH		
firm, in situ	50	(0.84:1)
loose ("vegetable soil")	28	(1.80:1)
SAND and CLAY	(37)	1.33:1
GRAVEL and CLAY	(37)	1.33:1
GRAVEL, SAND and CLAY	(33)	1.50:1
AVERAGE SOIL	(37)	1.33:1

Safe slope commonly assumed in practice (for average soils) is 1.5:1 to 2:1 (about 26°).

For granular soils a slope flatter than the slope of repose may be assumed as a safe slope without regard to height.

Slopes of cohesive material require flatter angles as the height is increased. This limiting height will vary as to the degree of compaction, compressive strength and angle of friction. It will also vary as to the foundation on which it rests.

AIR

Earth shelters—at least those that are well designed—are not dank, odiferous caves. They don't smell like musty basements. They don't become overly stuffy or oxygen-depleted. This fear is one of the commonest heard by earth shelter architects and engineers, but it is groundless. Earth sheltering presents unique ventilation and humidity problems, but they can be easily and confidently handled.

A surface building has an intense, ongoing relationship with its surrounding air. It trades stale inside air with fresh outside air through windows, pores in walls, cracks in floors and ceilings, and innumerable other pathways. In a surface dwelling the designer's problem is to lard enough caulking, insulation and building material between inside and out to hold air flow down to reasonable levels; otherwise the home's heating costs will go completely out of sight.

An earth shelter, on the other hand, is exposed to much less surrounding air and is far more carefully sealed than a typical surface house. It will not naturally trade air with ease. It has fewer open joints, less window area, more shelter from blowing winds, thicker walls, and dozens of impediments to air flow.

While in a surface home the problem is holding down air flow, in an earth shelter the concern is stimulating it. This situation presents both advantages and disadvantages. On the good side, the designer has far greater control in trying to build proper ventilation than he does while trying to prevent excess air leakage. Through clever design an earth shelter can be made into a perfect living environment—a splendid balance between fresh air and energy conservation. The earth-sheltered situation is more stable than that on the surface, so the designer can bring greater certainty and finesse to his ventilation plans.

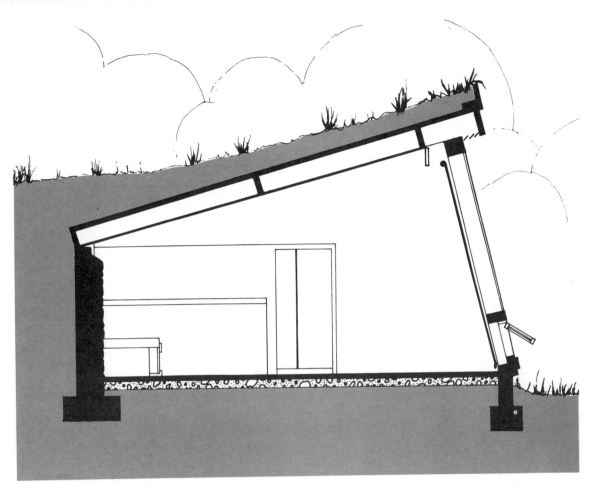

David Wright has devised an ingenious air-circulation system. In summer the shade covers the window and air enters the house through a vent under the roof. It slides along the roof to the back of the house and then out along the floor. In winter the shade is lifted and the upper vent is closed. Air enters at the window bottom and is trapped between the panes, where it is heated and rises. The hot air becomes free heat for the house.

ENSURING FRESH AIR

The drawback is that earth-sheltered ventilation requires greater care and concentration than a surface home. If a surface-home designer makes a little error in calculating air exchanges, the home will still be fresh enough; it may simply cost a bit more to heat. In below-grade situations greater caution must be taken to insure strong ventilation under all circumstances so that the house remains comfortable to live in.

Ventilation is required in all houses for a number of reasons. First is oxygen. Inhabitants need a certain amount of fresh air to breathe. Ventilation must also get old, stale odors out of the house so it doesn't become fetid. And it should trade house air that has been dampened by perspiration, cooking liquids, bath evaporation and such with drier outside air.

No specific test exists to determine precisely how much fresh air a house needs. But experience has shown architects that two complete exchanges of air a day represent a reasonable minimum. Less than this and the house grows uncomfortable; more than this wastes energy.

In earth shelters both passive and active systems can be used to generate the proper level of ventilation. Windows and skylights can be designed to open. They can be placed to maximize the flow of air. The key to such passive ventilation is the moving of air from one level to another. Either it must enter high and exit low or vice versa.

Air scoops and clerestory windows high in an earth shelter can suck in surface breezes and angle them down into the house. They will flow down and out toward an exit situated at a low level, say, facing a sunken courtyard or atrium.

By the same token, open skylights high in the ceiling of a submerged home will release rising, heated air to the sky. As this air heads out, it can draw in cooler air through a lower down opening, such as an open door, creating a natural airflow. If the skylight opens with a crank, the amount of ventilation can be precisely controlled by the homeowner, depending on the day's breeze and temperature. The cooling effects of such a ventilation plan can be sizeable.

The effectiveness of natural ventilators in an earth shelter can be enhanced by proper positioning and landscaping. Hills can direct breezes toward the house, as can strategically placed trees and shrubs. The house itself can aid ventilation by being placed at a 90° angle to prevailing winds. Even if the house can't be perfectly fitted to local breezes, the scoops and skylights can.

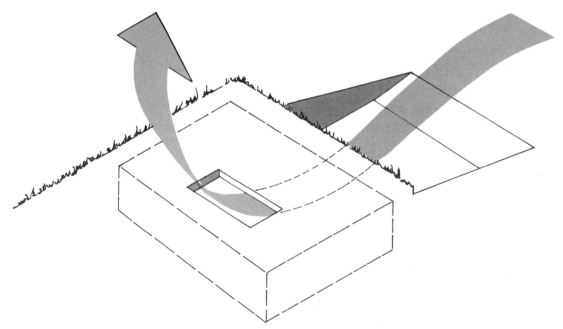

An open skylight creates airflow by giving heated interior air a place to escape. The rising air pulls in cooler air through lower openings.

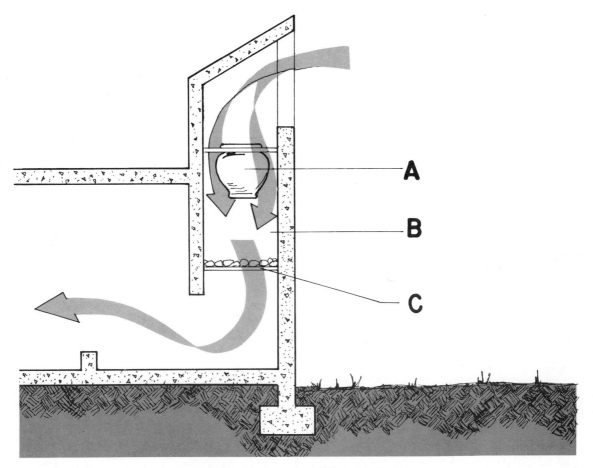

Free air conditioning from the desert. Placing an urn of cool water (A) and a screen full of charcoal (C) in the center of an air scoop (B), pre-chills and filters air as it enters the house. Nomads have used it for centuries and it would work here, too.

One additional consideration for earth shelter ventilation crops up when a fireplace or wood stove is included. Because these oxygen-hungry burners will stand below grade, they won't draw as well as they would in a surface home. The amount of oxygen they will suck out of the carefully controlled home environment could throw ventilation calculations way off. Generally this problem has been solved by running a small duct from the surface to the stove. The burner uses this separate air supply, leaving the house air fairly stable.

Active ventilation systems for below-grade houses include forced-air ducting or good old-fashioned window exhaust fans. Forced air ties ventilation in with the furnace that heats the house in winter. This furnace can be designed to provide the proper level of fresh air during the heating season by mixing returned heated air with outside air drawn through a vent to the surface. During summer the furnace fan can push air around the house using 100% outside breeze.

If the house isn't going to have a forced-air heating system, a window exhaust fan can create a breeze in summer. If the fan is set at one end of the house and a small window or skylight is open at the opposite end, a cross-house breeze will spring up. In wintertime a fan ventilation system might be able to take advantage of the heat-tempering properties of the soil. Instead of a window the fan could draw air from a duct that wended its way down from the surface, through the warm ground, before reaching the house. In this way the air is preheated before entering the house, reducing the heat loss caused by needed ventilation.

In hot climates the same process could provide precooled air for comfy summertime ventilation.

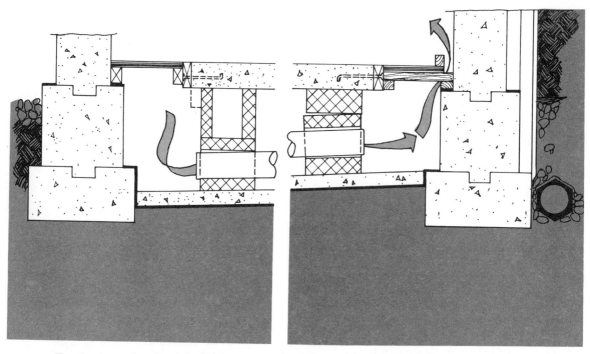

Earth pipes circulate air below an earth shelter floor. In summer the earth absorbs heat from the air before it enters the house. In winter frigid air can gain extra heat from the ground, thus cutting energy costs.

HUMIDITY CONTROL

Ventilation regulates humidity as well as temperature. As air stagnates in a house, it builds up moisture. In a surface house air doesn't hang around long enough to get seriously damp. But underground, if ventilation isn't adequate or doesn't reach all parts of the house, pockets of extremely humid air can pop up.

Not only does humid air present a comfort consideration, it also presents condensation and mildew problems. Condensation occurs when humid air comes in contact with a surface that is sufficiently cool. Two factors control the appearance of condensation: relative humidity and dew point. The relative humidity is the balance between temperature and the amount of moisture in the air. Air expands and contracts with changes in temperature. So if the amount of water in the air remains constant, the percentage of water relative to air in each square inch of space gets higher as the temperature rises and lower as the temperature drops. The higher the relative humidity, the more easily the air can dump its load of moisture. The temperature at which the air will drop its moisture on a solid surface is the dew point. The higher the relative humidity, the higher the dew point. When a surface drops below the dew-point temperature, humidity in the surrounding air will condense on it in the form of water drops. This is the reaction you see when water and fog form on the side of a window in wintertime. The outside weather drops it below the dew point and it pulls water from the air.

In an earth shelter condensation can be messy, ruinous to paint and wall coverings, and conducive to mildew and other molds. It can be prevented by either warming the surfaces or keeping the air from getting too humid.

Proper insulation and vapor barriers (see the section on Insulation) can help keep wall temperatures from dropping too low, and well-planned ventilation can keep the air moving steadily enough so that it doesn't pick up too much moisture. If a problem occurs in a particular room, Malcolm Wells recommends a small circulating fan to push the heavier damp air up and draw the lighter dry air down, creating a mild swirl that spreads the humidity more evenly through the air. In many cases, he says, this is enough to take care of the trouble. For more extreme cases, such as a house in a terribly wet climate, an electric dehumidifier or air conditioner might be required. Both will suck moisture from the air—at a price.

A final note about humidity: Since earth shelters are often made of concrete, which has a layer of waterproofing on the outside, moisture problems may be worst during the first few weeks of occupancy. Concrete is a great water retainer. Much of the moisture floating around during construction and the first days of occupancy soaks into the walls. It can't get out the other side because the moisture barrier keeps inside water in as well as outside water out. During early occupancy the walls and floor will carry an extra load of moisture that they will slowly release into the house. This may necessitate the presence of a dehumidifier until the walls have dried out.

Once the walls have released their initial load, however, their enormous moisture-gobbling properties will help keep the house air from becoming too uncomfortably damp. As with temperature, they will help keep house humidity on an even keel.

WATER

Water, both in the ground and on the surface, can plague an earth shelter. But many techniques in design and construction can handle just about anything water is likely to do. Many subsurface dwellers often worry that their homes will be as leaky and damp as surface basements. According to the American Underground Space Association, "Forming an opinion in this way is like looking at a barn and deciding houses shouldn't be built of wood above ground because the wind will blow right through."

The site is crucial in establishing a watertight earth shelter, because it determines the amount and direction of surface flow as well as the moisture content down below. When judging a site, start with the obvious—floodplains and plots with very high water tables don't make much sense for subsurface living. In marginal sites a bermed design, which stands at grade rather than below, can alleviate some, but not all, water worries.

The best earth-sheltered site is in a gently sloping area—between 1% and 5%, according to the American Underground Space Association—free from gullies, washes, flash flood sluices or severe erosion. An extremely steep slope will tilt water away from the house but will make it hard to maintain gardens and rooftop vegetation; it also means that water from above can plow toward the houses under great pressure.

BASIC TERMS AND PROPERTIES

Before you build an earth shelter, there are some basic water properties and terms you should be familiar with:

Percolation refers to the slow seepage of water downward through the soil. The slower the water percolates downward, the more moisture gets picked up by the surrounding soil and the wetter the environment in which the subsurface structure rests.

The water percolates through the ground until it reaches the *water table*, which is the level at which the soil becomes saturated. This is why postholes and excavations fill with water at the bottom. The water table isn't a fixed distance from the surface; it varies with rainfall, runoff and the terrain. But since water tends to level out, the overall water table in a given area is fairly predictable.

Capillary action is a phenomenon by which water climbs upward or sideways through porous material. So the area just above the water table is filled with percolating water from above and capillary-drawn water from below. The walls of an earth shelter create a capillary attraction with the surrounding groundwater. They soak up the moisture like a sponge. To prevent the walls from soaking requires waterproofing barriers and capillary breaks. The barriers consist of sheets of material laid over the outside of the walls. The breaks are stretches of fast draining material that eliminates capillary draw.

A much-touted system for breaking capillary draw comes from Sweden. It is generally considered to be the best possible creeping-water eliminator, except in the most extreme circumstances, and it's as simple as can be. The Swedes had the notion of putting about two inches of fluffy rock wool sheets between the gravel backfill and the house walls. The insulation is held in place by its rigid backing, but the interior is so airy that the water can't get from strand to strand; it is literally stranded before it reaches the wall. The Swedish break reduces water pressure against a wall and can reduce, or even replace, the need for some of the more complex waterproofing methods to be discussed shortly.

The one drawback to the system is that it falls apart if more water builds up outside the insulation than the drains can haul away. If the spaces between the insulation fibers fill with backup water, the effectiveness of the capillary break goes from 100% to zero. The Swedes found that some protection in case water pressure against the wall rises can be obtained by coating the wall with two coats of asphalt or waterproofing to a height of twenty inches above the footing.

Hydrostatics means water pressure. Groundwater generates a load called hydrostatic head, which gets greater the farther the water stands from the water table. This works both above and below the water line, creating downward pressure from above, upward lift from below, and horizontal pressure in all cases. You can gain a sense of the pressure water brings to bear on a building from beneath by pushing a rubber ball down in a partially filled bathtub. The water forces the ball back to the surface. Groundwater tries to do the same thing to your house.

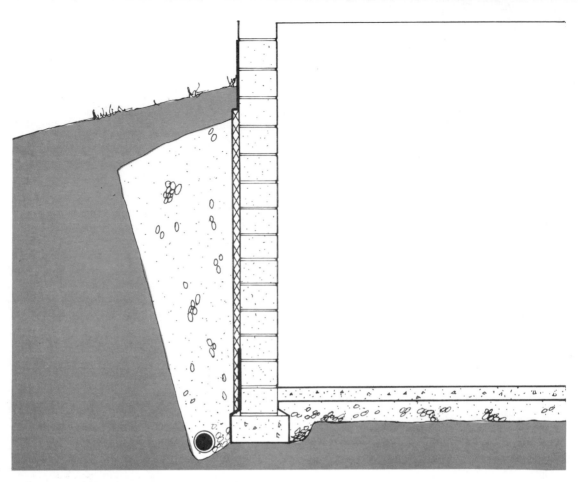

A Swedish break. The crosshatched layer is rock wool, which stops capillary action. Water can't creep through it to the walls.

DRAINAGE

Many devices have been invented to deal with the problems and pressure associated with groundwater. It is entirely possible to build an earth shelter that extends well below the water table. But there are dangers, and Malcolm Wells, for one, feels they aren't worth the risk. "Although some architects do it, we say don't build below groundwater levels. Sooner or later the drains or the waterproofing or the sump pumps will fail, and what a mess! But if the groundwater level is high, don't despair. Many underground buildings are built above grade, with earth piled up along their walls and mounted on their roofs, creating, in effect, artificial hills."

Waterproofing will be necessary in any earth shelter, but all waterproofing systems will work better if prudent designs for drainage have been included in the plans. Water has the pleasing characteristic of laziness. If you give it an easy way out, it will take it. Drainage systems are predicated on the principle of giving water an easier choice than flowing toward—and eventually into—your house.

On the surface the most obvious form of drainage is conscious sloping. If the surrounding soil is lower than your house, surface water will run away from the building. Water on higher ground can be induced to flow around your building by swaling (i.e., the contouring of the slope into a curve that angles away from the structure). In many cases surface sloping can be augmented by shallow drainpipe arrangements that siphon off the running water to avoid standing pools.

Beneath the surface the same basic idea takes more complicated forms. When an earth shelter is built, it is placed in the center of a large hole. When the house is finished, the hole is filled up. Well-planned subsurface buildings don't fill in all that space with dirt. Instead they envelop the structure with a backfill of sand, gravel or other porous material. Water heading toward the house plunges through these fast-draining stones toward the water table.

At the bottom of the backfill field a good designer will place drainage tiles. These are perforated pieces of pipe that are laid around the foundation and are tilted slightly downhill. Draining water enters the pipe and runs away from the house. In locations with really bad drainage the tiles can be extended well away from the house. Or they can run to a storm sewer, or even a sump pump, to get the water out of the house environs.

Sump pumps are also necessary when the house lies below the water table.

Interior and exterior drains laid alongside the walls of an earth-sheltered home under construction. They'll keep harmful moisture from building up.

Water gravity drains to its prevailing level (the water table), but once there it sits uncooperatively. The only way to get rid of it is to drag it off with pump power.

Generally, though, earth shelter designers highly recommend gravity drainage over mechanical means. Unlike an electric pump, gravity won't ever conk out during a rainstorm.

When an earth-sheltered house lies right along the water line, extra drains may be needed to keep the house dry. Drains only bring water levels right down to the table in the space immediately above them. On either side the level of water in the ground curves upward, the height of the curve being determined by the properties of the soil. So it could be possible to have drains at either end of your house yet have the natural water level rise between the drains to the middle of the wall. Extra drains placed under the floor can flatten out the curve and keep the house dry.

At a hilly site, where there is significant surface runoff, or in a location with lousy natural drainage, French drains spaced about the site can nip the drainage problems in the bud. A French drain is simply a long trench angled downhill. It is filled with gravel or coarse sand and covered with a thin layer of soil. Water filters down through the stone to tiles that carry it away. It's like an express lane for water, with no stops near your house.

Let's say a site is ideal except for an area at the base of a steep rise where rainwater builds up and eventually comes charging toward the house's location. A French drain placed there could divert the water and drain it around to the side of the house, where gravity could take over and drag it away.

Water on the roof of an earth shelter presents special problems of its own. If the roof covering is flat, water can pond (i.e., stand still and build up) in spots above the living area. As you can imagine, that situation could be deadly for subsurface comfort. Happily, plant life on the roof can often take care of roof drainage problems. The roots serve as nature's own drain tiles. In other cases a slight tilt to the roof can drain off enough water to eliminate worry.

If water is tilted off a roof, however, it has to go somewhere. You don't want a sheet of water cascading over the roof edge, any more than you would in a surface home. Gutters of some sort might work, but earth-sheltered designers seem to favor rooftop drains that drop the water down pipes hidden by interior walls and partitions. They don't ruin the line of the roof (a consideration with earth shelter houses, where the roof line lies at eye level or below, not up above the viewer's head) and link well with below-surface drainage paths.

Performance has shown that a well-designed earth-covered roof can be just as dry and puddle-free as the angled crown of a Cape Cod. An engineer at Terraset, the earth-sheltered school in Virginia, for instance, crowed that "it's the only roof we have in the school district that hasn't had a leak."

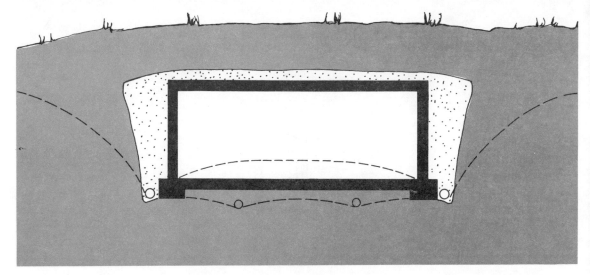

An earth shelter is a sealed box in the middle of a moist environment. Sandy backfill takes care of some groundwater, while drains beneath the floor handle more. Sometimes drains along the sides suffice, but in other circumstances drains under the floors are required because water levels rise on either side of a drain.

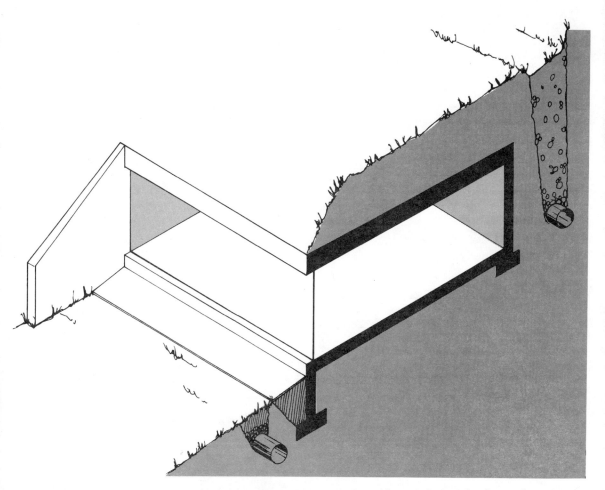

Drainage at an earth-sheltered house site: a swale and gravel trench above the structure, a shallow drain below and a sloped roof cover.

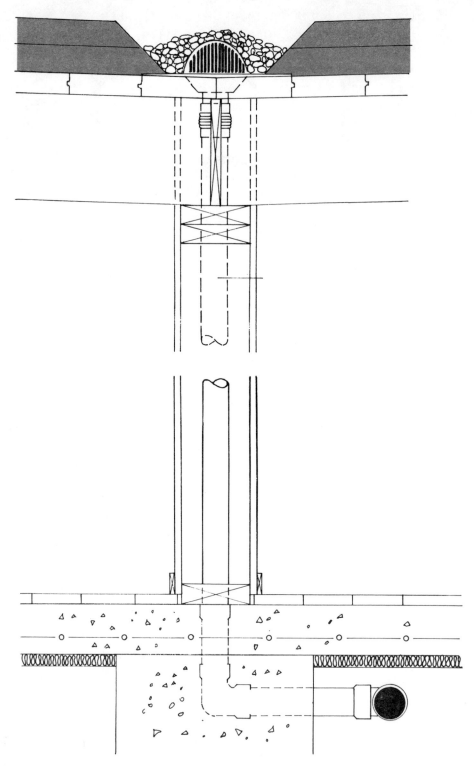

The preferred method of draining water from an earth shelter roof is to place a drain above an interior wall or column and run a pipe down to the subsurface drains.

WATERPROOFING

Even with the best design, an earth-sheltered structure will still require careful waterproofing. The most extensive waterproofing arrangement might not be absolutely necessary, but when underestimation means a wet mess and digging up the yard, it is better to be safe than sorry.

Waterproofing can be attacked from two directions: internal and external. Smart earth shelter designers use both. Internal waterproofing involves making the structural materials themselves as waterproof as possible; external waterproofing involves sealing them off from moist surroundings. Internal waterproofing works best at keeping out water vapor; external waterproofing holds out water itself. Concrete, with its extreme thickness and density, is relatively waterproof all by itself. Many chemicals and coatings can be used to make it still more waterproof. Some of the commonest are:

Pargeting. Dense cementing material like plaster or stucco is trowelled onto the outside surface of the concrete wall. The denser material fills in cracks, gaps and minute spaces in the concrete.

Additives. Some concrete blends have chemicals already in them to limit shrinkage and slow down seepage. Using a concrete with such additives is a good idea. Adding waterproofing compounds to regular concrete is also wise. But by itself concrete mixing won't be dependable enough underground.

Asphalt. Sprayed or brushed onto the concrete, this sticky, smelly substance repels water. Sometimes it is applied hot, which makes for a thicker coat; sometimes cold, which fills in cracks a bit better. It doesn't have a long effective life-span, however.

Pitch. This is the sticky stuff used on roofs. It works reasonably well below grade, but it has drawbacks. On the roof the sun regularly softens the stiff material, letting it flow into cracks and remold to the surface. Below ground the cool environment will keep the pitch stiff once it has been applied, limiting its waterproofing ability.

If you get the impression that none of these substances gets rave reviews from earth shelter engineers, you're right. They help limit water exposure but they themselves can't stop it for good. The biggest drawback of all of them is that they are part of the wall. When the concrete expands and contracts, so do the seals. Eventually all this motion will crack and loosen them, just as it does to the concrete they are attached to.

In time they will simply wear out. Above ground this wouldn't be that big a problem; one could simply apply a fresh coat. But you don't want to have to dig your house up to reseal it every ten years.

So earth shelter designers invariably buttress the internal waterproofing methods with external materials. The combination of coatings suffices to keep the water out without worry. Here are some common systems of external waterproofing:

Built-up Membrane. First a thin layer of waterproof material—polyethylene for instance—is stretched over the outside of the walls. This would suffice to keep out water, but backfill tears, soil acidity, roots or frost could break it and ruin the protection. So some designers add on layers of reinforcement such as felt, burlap, fiber glass or canvas. This is covered by a second layer, or sealant, and so on until a sufficient protective thickness is reached.

The architectural guidebook *Time-Saver Standards* lists the following recommendations for layers under different situations:

WATER HEAD (IN FT.)	COAL TAR/FELT	ASPHALT/FELT
0	2	2
1–3	3	3
4–10	4	4
11–25	5	5

Polyethylene Alone. It's cheap and you don't have to worry about sunlight damage (the material's biggest drawback) underground. It may not be adequate alone to face up to the water pressure on earth-sheltered walls, but the American Underground Space Association feels, "for the floor of an earth-sheltered house that is above the water table and has a foundation drain system and gravel layer under the floor, this type of damp-proofing protection will usually be adequate." Sometimes polyethylene can be embedded in a coating of mastic; this provides better protection than the plastic by itself but still doesn't hold up well under steady pressure.

Bituthene. This is polyethylene that has been coated with rubberized asphalt. It offers some of the advantages of a membrane with the convenience of polyethylene. It stretches extremely well. Bituthene is hard to seal if the wall is at all damp or cold. It has the disadvantages of polyethylene—it can't stand up to constant heavy pressure—and membranes—its leaks are hard to pinpoint and the layering sometimes imperfect.

Butyl Rubber, EPDM, Neoprene Compounds. These are plasticlike substances that can be bought in long, thin sheets. Malcolm Wells uses these. They get high marks, with the possible exception of sealing along the seams. One way to reduce this problem, which can result in water running between sheets of waterproofing, is to glue the material down in a tightly patterned grid. Then even if water gets in it will be trapped in a chamber bounded by glue lines on four sides; it can't run amok between the sheets. The quality of these sheets seems to vary enormously from company to company, so check with an engineer before buying any.

Liquid Sealants. Polymer plastics in liquid form can be sprayed or trowelled onto a wall. Generally, if you can use a different sealing form do so. But in tight corners or uneven spots where sheets won't work, this provides a passable alternative.

Bentonite. This clay expands on contact with moisture. Sandwiched between cardboard panels, it provides an antiwater layer that gets stronger as the

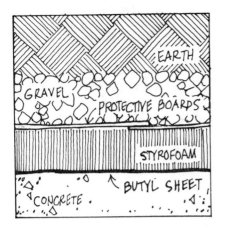

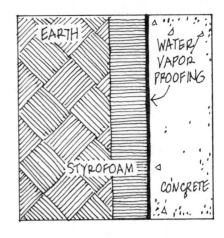

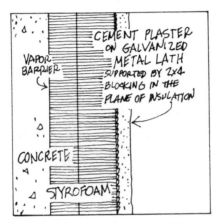

Three waterproofing ideas. Above left, a typical rooftop system; above right, the simplest wall waterproofing method; below center, a more complicated system with additional waterproofing on the inside surface of the walls.

leakage gets worse. It can be put up more quickly than other sealants and has been used successfully in a number of earth shelters, despite some questions about how well it reseals after expanding. Bentonite doesn't degrade, but running water can erode it, so it shouldn't be used in the juiciest soils. Nor does it work well in brackish soils, because the salt blocks the necessary swelling. Spray-on bentonite is also available.

Trouble spots for earth shelter waterproofing are corners and around openings and projections, such as skylights. Corners must have carefully overlapped and sealed protection. As Malcolm Wells pictures it in the diagram, the sheets form an angled T in which one wraps around the outside of the insulation sheet on the other wall. This eliminates any straight seams through which water can squeeze itself. With the overlapping, the water has to make a hard right and a hard left turn to find any exposed wall—more trouble than it is likely to take.

As for projections, some designers coat their attachments and immediate surroundings with rubber cement, which repels water. In addition, waterproofing material should be thickly layered around them. If possible, the waterproofing should enclose any exposed edges on the wall-break. The waterproofing should go over the skylight, not the other way around. Projections set in at a slight angle have less water-related trouble than those that are flat, since the moving water doesn't have as much time to find an opening as with the standing stuff.

Costs for different waterproofing systems vary quite a bit. Here, as a general guide (local conditions may affect some prices) are typical costs as of the end of 1980:

TYPE OF SYSTEM	APPROXIMATE COST/SQ. FT.
Internal	
Ironite pargeting	75¢–$1.10
Asphalt or pitch coatings	
brushed on	20¢–35¢
sprayed on	22¢–25¢
trowelled on	30¢–45¢
addition coat-add	20¢
External	
Polyethylene sheet	
.010-inch thick	13¢
.004-inch thick	8¢
Built up roofing w/fabric	
1 ply	35¢–55¢
3 ply	65¢–90¢
5 ply	85¢–$1.35
Bituthene	80¢–81¢
Butyl, EPDM, neoprene membranes	$1.50–$2
Liquid membranes	70¢–85¢
Bentonite	
panels	60¢–$1
trowel on	40¢–80¢

Making a waterproofing choice for an earth shelter has to be a bit of a gamble because none of the houses have been around long enough to provide certain long-term data about performance. We simply don't know which system will perform best in fifty or a hundred years. We do know, though, that care in attaching the waterproofing to the structure is an absolute necessity that, if ignored, will negate the worth of any system in a matter of weeks or months. This is literally one area of earth shelter building where you can't cut corners.

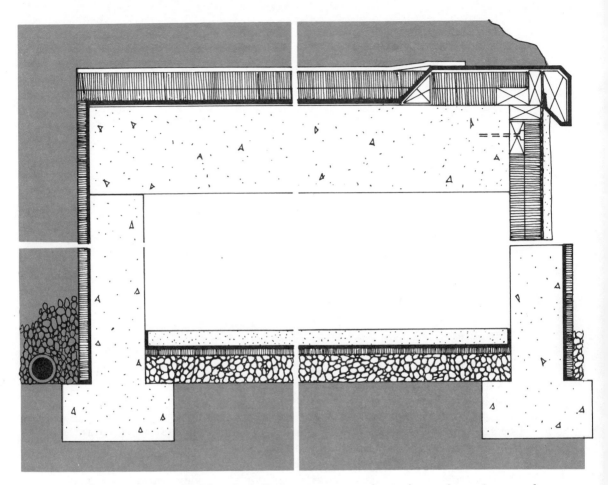

Waterproofing materials (dark line) must cover the entire surface of an earth shelter and must overlap at corners and outcroppings.

STRUCTURE

Far more so than in a surface building, the structural needs of an earth shelter control how it looks and feels to live in. The weight of earth and of the materials needed to support it have as much to do with the design of a subsurface home as any artistic designs of an architect or builder.

That's not to say that great grace and beauty aren't possible through earth shelter design; quite the contrary, these homes can be striking, with a natural allure and balance no redwood-sided surface home could attain. But structural integrity must come first and looks second when you're working with the weights and problems involved here.

In days to come, as earth shelter grows more common and the thought of living beneath a layer of soil becomes less unusual, radical new ways of solving the structural dilemmas of the earth will certainly come to the fore. We may see domed or geodesic earth-sheltered homes as the norm. Homes may be tube-shaped or curved collections of computer generated sections. For now, however, the goal of most designers is to solve structural problems with conventional means to produce a house as much like a surface dwelling as possible. In these early years of earth sheltering it is important to show how normal and liveable underground space can be, not how wild and unusual.

So some of the accepted structural solutions in today's homes grow from a desire for maximum normalcy rather than maximum efficiency. The objective is to retain the general look and sensation of surface living, and the methods employed today achieve that goal.

Coping with the design problems of earth shelter can result in alluring forms that bear no resemblance to cave openings.

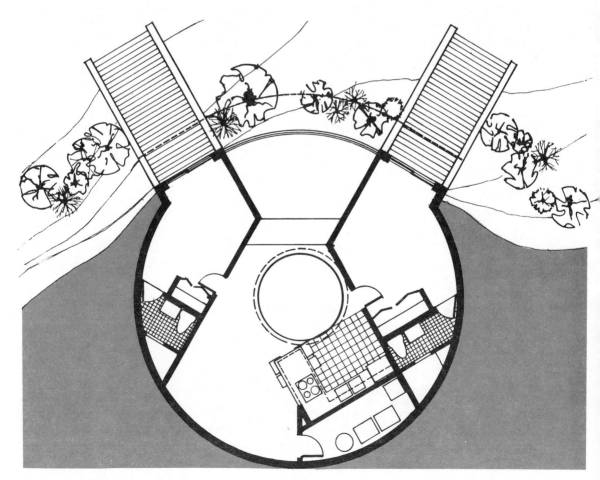

Earth shelters don't need to be rectangular or square. (This round design is by an Ohio firm.) But most earth shelters still conform to the shapes we are used to from existing buildings.

BASIC TYPES

The first major consideration in the structure of an earth shelter is its relationship to the surface. Both the depth of the house and its orientation affect this relationship. Beginning with the most surface-oriented and working our way downward, these are the major forms of earth-building patterns:

Berm and Roof. Earth rises to the height of the windows and covers a roof that sits like a terratecture cap atop the light-filled airy interior. The earth-covered roof is caplike because it never touches the soil along the sides. This is basically a surface home that adopts earth-sheltered ideas. It is the most accessible, has the most flexibility of future design changes, the best range of outside view, and the easiest access. Energy savings are slashed, however.

Courtyard. Imagine a surface house formed around a central courtyard. Now cover the walls to roof level with earth. Three walls of each room are sheltered by the soil, with the fourth wall facing the open atrium. Here again light reaches many areas of the house easily. The atrium offers the ultimate in patio quiet and privacy. The sense of being soil-covered is slight. However, all views are of the atrium, not the surroundings. Also, the atrium might be quite windy if not properly designed. Expansion of the house would be difficult. Entry positioning might be a problem.

Elevational. Earth is bermed up over the back onto the roof and rises on the sides to the height of the windows. The difference between elevational design and berming is that an elevational home is totally buried from the roof edge all the way down the back wall; a bermed house roof never touches the side soil cover. Basically you are creating a hill above grade in which to couch the house. Your view is toward the front, with limited exposure through the bermed sides of the house. Energy savings are relatively high and passive solar adaptability is superb. Design and appearance will be markedly different from a non-earth-covered dwelling.

Chambered. The house is completely enclosed by the earth except for a limited means of entrance and exit. The surface world is replaced by electronic simulations of light, wind and environment. Energy savings and security are maximum, but so is the alien nature of the living experience.

Excavated and Topped. Instead of bringing the earth up, as in the bermed house, you drop the house down. Excavation sets the house about roof level beneath grade. A roof covered with soil is set atop windows, which cover all four sides of the house. The rising land outside the house limits the view, and its lowness may limit solar potential. The house is protected from surface winds, which increases energy savings, as does its lower position in the soil.

Inset. Where a natural slope exists, the home can be excavated back into the hill rather than downward. This has great energy-saving potential. It also puts the house on the cheaper land, leaving the better land free for a view or recreational use. Getting enough light throughout the house is a concern.

Atrium.   This style is like the courtyard design, only sunk into the earth rather than bermed. It has the same problems and positive points. In addition, access would have to be through the atrium, which requires climbing a flight of steps.

Vault.   This type is similar to the chambered design, but below grade. The feeling of being in an enclosed space would present the same challenge.

Two-story, Above-below.   A second story can be added on to an earth-sheltered house that rises above grade for better viewing and lighting. Some loss in energy savings is inevitable, however.

Two-story, Below.   An elevational or other near-surface design can be linked to a vault or chamber, creating a second story that is fully enclosed by earth. This increases the ratio of interior space to surface area, which makes energy savings much greater. Use of such rooms as offices, storage spaces or bedrooms might reduce the enclosed feeling they could have.

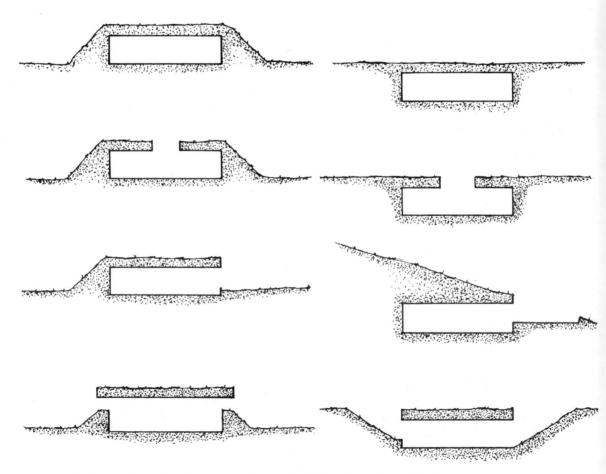

Basic earth shelter forms are determined by the relationship between structure and surroundings. Left column (from top): chambered, courtyard, elevational, berm and roof. Right column: vault, atrium, inset, excavation and top.

The weights and stresses of building with earth require a uniquely sturdy and carefully constructed foundation. Concrete is the usual choice.

STRUCTURAL FACTORS

Beginning from the bottom up, here are some of the important structural factors that control earth-sheltered designs:

Foundations. Because it has so much weight pressing on it, an earth-sheltered building requires a sturdy, carefully planned foundation. The footing is the bottommost part of the foundation, the anchor for the whole house. It needs to be big enough to support the structure against settling into the soil or shifting horizontally. Malcolm Wells prefers a tiered footing, with a broad, flat base and a narrower, taller second stage that joins the wall and floor.

Outside, the foundation of an earth shelter consists of a bed of stones, which enclose drain tiles (more details on all of this may be found in the section on Water) that carry moisture away from the building. Between this drainage area and the wall stands a layer of insulation board over a layer of waterproofing material. The insulation and waterproofing also rest under the floor. Straw is spread above the stones, with a backfill of site soil covering the distance to the surface.

In some cases the base of the wall may be cantilevered to give the house still greater stability relative to the surrounding soil. A cantilever is a horizontal flange that sticks out from the base of the wall just above the footing. It makes the bottom joints look like inverted T's. Cantilevers allow the wall to accept more weight from above.

A few engineers raise the possibility of using pressure-treated lumber for earth-sheltered foundations. The main advantage here is ease of construction—any home handyman is familiar with wood construction. Wood is also lighter and far easier to work with than concrete. The big—and, most architects feel, fatal—drawback is that wood has none of the thermal advantages of concrete.

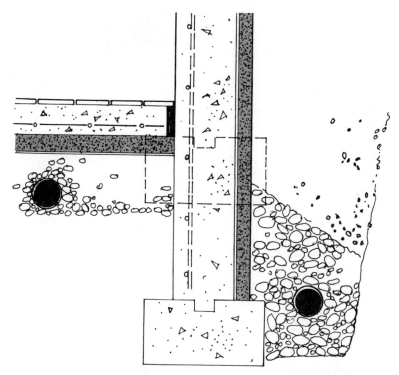

Typical foundation corner for an earth shelter. The concrete wall, with its core of reinforcement bar (dotted vertical lines), sits atop a thicker concrete footing. Waterproofing and insulation cover the outside surfaces, and gravel-covered drains rest just below and beyond the building's edges.

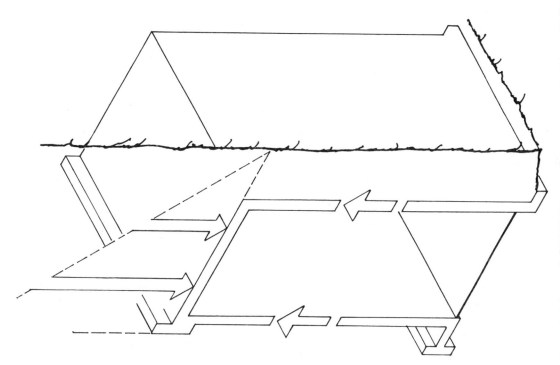

The stresses on earth shelter walls come from the soil and the rest of the building. The inverted T-shapes at the base of these walls are cantilevers, which increase strength. Stresses are passed along the building surface to the walls best able to support them.

It doesn't store or transmit heat at all well. And since heat interaction and thermal mass are the major energy-saving features of earth sheltering, they shouldn't be squandered for convenience.

Walls in an earth-sheltered design are made of poured concrete or concrete blocks, with poured concrete by far the most common. These walls are formed by being poured into a form, just as in the creation of a surface basement. The forms invariably contain lengths of thick, round iron called re- (for reinforcement) bars. This strengthens the concrete, ties it together under tension, and provides a heat transfer medium to reduce cracking from expansion and contraction.

Concrete is used for soil-covered buildings because it can accept enormous loads—and gets stronger as it gets older. Within the cocoon of earth, the thermal mass of concrete is also a positive feature. Thermal mass refers to a substance's tendency to absorb, retain and transmit heat; the higher the thermal mass, the greater the moderating effect of the substance on the surrounding environment. However, when it comes in contact with the surface, concrete's high thermal load can be a detriment. The wall will give up all of its retained heat to the cold out-of-doors, ruining the energy-saving belowground situation.

To alleviate this problem, thermal breaks—strips of insulation the width of the wall—are laid in with the concrete at strategic points. These inserts stop the flow of heat to places where the concrete touches the outside environment. They are crucial for a smoothly energy-efficient earth shelter.

Sewer and water pipes, gas lines, window frames and electrical conduits are also placed into the concrete wall when they are poured. In this way they can be instantly insulated by the walls—no frozen pipes in winter—and hidden from view.

Precast concrete slabs can also be used in earth shelters, but, frankly, they are more trouble than they are worth. These factory-made concrete panels are expensive, difficult to transport to building sites, and not particularly labor-saving. The joints between slabs provide excellent routes for water leaks and you can't get pipe and conduit passages exactly where you want them. Pouring is easier and offers more control.

As for concrete blocks, Roby Roy, a self-styled earth-sheltered designer, built his New York home using a newfangled method known as surface bonding. This technique avoids the use of mortar—the weak cement that binds bricks in surface houses—which would never survive the rigors of life in the soil. Instead of being joined by mortar, the blocks are coated with a special parget. (See the section on Waterproofing for more details.) This coat of cement and sealant, according to government and university sources, produces a wall six times as strong as one using mortar. It could be strong enough for in-soil use.

Re-bar can be used to strengthen blocks still further. Without re-bar a one-foot-thick block wall can safely rise eighteen feet; with re-bar it can go to twenty-five feet. More information on surface-bonding block walls can be obtained from the U. S. Department of Agriculture.

One other thought about blocks: They might work particularly well for interior walls, where the worries about water pressure and load aren't as great as along the edge. Even interior load-bearing walls could be made of blocks. They will be far stronger than wooden partitions and will enhance the home's thermal potential. All interior block walls must be anchored to the exterior walls using I-shaped metal tie bars.

ROOFS

The roof is the focus of an earth-sheltered design. It generates the biggest load and provides balance and stability for the house. It ties all the other structural elements together.

Opinions vary among designers as to the best way to make an earth-sheltered roof. Roofs of earth-sheltered homes can be made of concrete or very heavy timbers. The commonest roof form is cast concrete. This offers the best weight acceptance and thermal properties, as it does in the walls. Pouring a concrete roof, however, is a bit tricky. The molding forms must bear all the weight of the wet concrete, so they must be shored up from below. This can be a slow and somewhat expensive procedure.

To get around some of these problems, John Barnard used precast concrete planks for the roof of his Ecology House in Marston Mills, Massachusetts. They saved money and worked fine. But other designers express grave doubts about precasting roofs. Andy Davis, of Davis Cave Homes in Armington, Illinois, fears that the weight of rooftop soil will eventually cause the planks to buckle along the seams, creating leaks and heat loss. Don Metz, an independent architect from New Hampshire, feels that only custom-made planks present any positive features, but even they have so many drawbacks that he couldn't see why anyone would put up with them. Metz favors heavy timbers for his roofs. He feels they provide sufficient strength, with greater ease and better looks than concrete. Instead of a roof that looks like a basement, Metz says, you end up with one that looks warm and richly rustic.

Whatever material is used, an earth shelter roof should never be absolutely flat. It must be slightly cambered, which means it curves downward slightly from the center line toward the edges. When heavy weight is applied to a cambered roof, it will settle into a flat formation; when weight is added to a flat roof, it bows downward, creating a drainage problem and a structural nightmare. So a slight bow to the roof is necessary.

A few designers have taken the idea of a bowed roof much further. In River Falls, Wisconsin, designer Michael McGuire created the Pat Clark house from half of a giant curved culvert. The curved configuration spreads the load of earth cover evenly through the structure. Similarly, some soil-covered buildings have been topped with geodesic domes, which distribute the weight far better than flat roofs. Geodesics, however, also dictate the shape of the structure, limiting the designer enormously.

A highly curved roof will always be able to accept proportionally more weight than a slightly bowed one of the same composition and size. As the bow

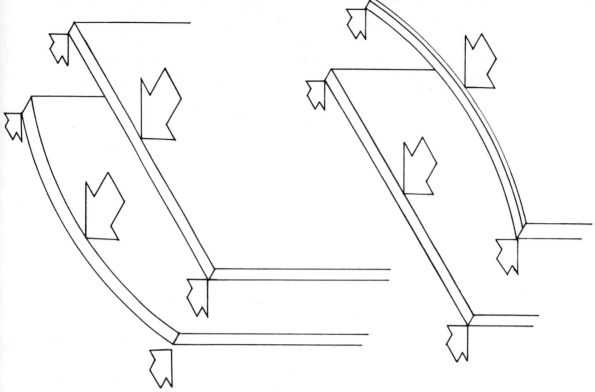

Earth shelter roofs should never be designed flat. A flat roof will curve down-ward under weight. A slight bow will flatten.

rises, the weight will press more steadily downward against the walls rather than outward; the roof has less opportunity to collapse and so can take on more poundage.

The central factor in roof design, of course, is the load. A surface roof must bear the load of its own weight plus any rain, snow or heavy winds that might press against it. An earth-sheltered roof will be made of heavier materials and will have to cope with the extreme situations of sopping wet soil or huge wet snow drifts that could vastly increase the downward pressure.

How much load might be involved? "It can range from 150 pounds per square inch," says Malcolm Wells, "for a roof with a wildflower ground cover to over 400 pounds per square foot when enough earth is used on the roof to support small trees. [For more on soil depth see the section on Soil.] In addition, a building code-prescribed snow-load allowance is usually added, and if the roof is used as a yard or terrace, a code-prescribed 'pedestrian live load' is involved, too. Designing for all these conditions, which will vary according to your particular needs, is within the scope of the most basic engineering practice. In other words your architect and his engineers are perfectly capable of designing such roofs."

For safety's sake the designer will assume the worst situation and create a roof strong enough to withstand it. He or she will examine live loads—people, cars, a herd of cows galloping over the roof—that come and go. The designer will also factor in dead loads: the combined weight of the roof components, soil, and ground cover pressing down on the walls and floor.

The reason a professional should handle these calculations is that the forces placed on an earth-sheltered roof can be extreme and hard to gauge. A tree, for instance, imbedded in roof soil and lashed by a stormy wind will push down on one side and pull up on the other, stretching the roof like a giant crowbar. How much total stress will that create? What about workers and machines that might need to be on the roof during construction? If bulldozers will have to traverse the top, their weight will have to be figured into potential loads.

As the load grows greater, the need for supports from below to keep the roof stable grows concomitantly. Columns can provide such support, as can cleverly designed interior walls. To some extent, the number and placement of interior walls in an earth shelter will be determined by the load on the roof. A deeply buried house may have to make do with smaller rooms, so that the load is evenly distributed throughout the structure.

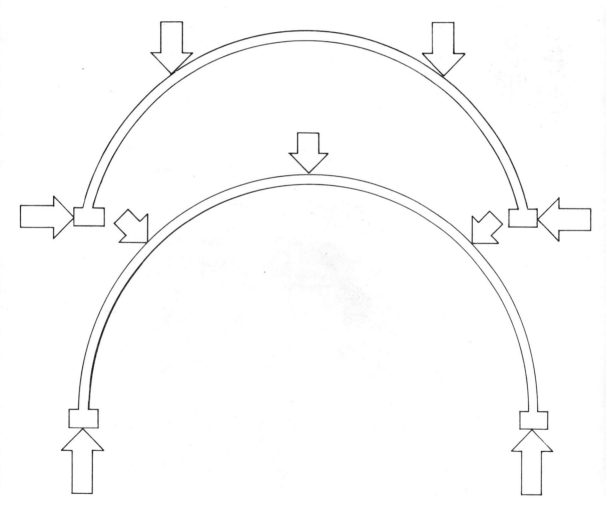

The height of an arch determines its load-bearing capacity. If the arch is wider than it is high, weight will press horizontally. If it is taller, the weight will push vertically down onto the walls.

Load problems can also be handled by keeping exterior walls short. One way to accomplish this is to build something other than a rectangular house. Andy Davis's cave homes are octagonal. William Morgan's dune house is completely curved. The Karsky house in St. Croix, Wisconsin, consists of a collection of interlocking semicircles that pass the load back and forth. By having more angles in the walls, each wall span can be shorter; since the many short exterior walls share the load better than a few long and straight walls, the number of interior load-bearing supports can be reduced, opening up the house to more design possibilities.

In a structure where all the walls are load-bearing and are tied to both the roof and floor, they serve as a mutual support system. Each part of the system puts tension on its attachments, thereby strengthening them. The whole is far stronger than the sum of its parts. However, when the loads can't be equally shared, as in elevational or inset designs, a different principle must be put to work. In such structures the south-facing glass wall is fairly useless for accepting loads. The roof then serves as a means to tie just three walls together: two sides and the back. This type of roof called a diaphragm, is stretched over the space enclosed by the three walls instead of sitting on top of the four-wall space of a tension system. A diaphragm roof must be designed differently than a tension-inducing one. It may have to be bowed so that the outward pressure is less, and more angled, and the weight presses down less directly on the walls.

Cleverly designed buttresses can take some of the roof's weight without inflating expense or cutting down on interior light.

The considerations that can go into an earth-sheltered roof can multiply indefinitely. But some of the more important and striking ones are listed in the following miscellany:

Parapets.   A parapet is a vertical ridge that extends upward from the edge of the roof to contain the rooftop soil. It is a bad idea. Malcolm Wells warns: "Look at any masonry building having parapets and see the deterioration up there, particularly near the corners of the building. When earth cover is added, the problems get worse, for a new factor—that of ice pressures—is added to the rooftop walls. As soil freezes it expands, and the expansion can damage—even overturn—such walls. Parapets tend to help trap water within the rooftop soil, too, so there's more likelihood of ice expansion in such locations. They also provide an excellent escape route for house heat."

Overhangs.   Passive solar designs call for roof overhangs calibrated to allow the lower winter sun to reach windows but not the higher summer sun. Some designers simply extend the roof material out far enough to provide the necessary overhang. They use thermal breaks set into the roofing material to keep heat loss through these extensions to a minimum. Malcolm Wells, though, feels such roof ledges cause "energy nosebleeds" from the house. He recommends solar awnings, which can be unrolled in summer and retracted in winter, instead of an extended roof. Other possibilities include adding a broad lattice or trellis around the roof. Vines on this matrix could shade the windows in summer but let the winter sun through their bare branches. Thermal breaks in the connections between roof and trellis could hold down energy loss.

Erosion.   Rooftop soil won't erode significantly if shored up by a surrounding layer of hay, mulch or other anchoring consisting of natural material. Malcolm Wells has found that even without the barrier, roof soil raked back from the edge at its natural angle of repose will not wear away. A solid barrier, like a plank railing, to hold the roof soil is not generally recommended because it hides the view of the roof plantings from the people below.

Railings.   Some sort of see-through railing, however, might be a good idea. Wells notes that "when roofs reach the ground, children reach the roof." Something to keep the careless or unaware from tumbling off the roof is essential. Bushes too thick to walk through could offer a natural solution.

Roots.   Despite the worries of potential earth shelter homeowners, tree roots don't present a danger to the roof or walls, assuming the structure has good drainage. Roots follow water, and if the water is effectively separated from the structure, so will the roots.

One further thought on structure. Because earth shelters seldom have basements, the mechanical equipment usually stashed downstairs will have to be placed closer to the living area than normal. In the quieter-than-surface environment, the noises of furnace, water heater, blowers, and so forth, can be most disquieting. Fitting these noisy units effectively into a structurally sound design is one of the principal challenges of earth-sheltered buildings.

The type of vegetation that can be grown successfully and safely on earth shelter roofs varies with the depth of the soil. At the left side of this illustration, only grass would work well; grass requires about a foot of soil. At a depth of eighteen inches, small shrubs become workable. At about two feet, large bushes can be used, and at three feet (the right side of the illustration), heavier trees will be supported. The rule of thumb is that there should be at least twelve inches of earth beneath the root ball of the plant.

PROFESSIONAL HELP

Finally, we come to the question of how much of the responsibility for planning an earth-sheltered structure should devolve upon professionals. A couple of books have been published recently extolling the virtues of the owner-built earth shelter. But the opinions of architects and engineers in the field (who obviously have some vested interest in getting people to use their services) range from mildly to violently negative on this important question.

"I think it's entirely an individual matter," says Don Metz. "I'm a builder myself and I've built a number of houses, so building a house to me is nothing to get excited about. One of my clients built his own house and did a real nice job. It just depends who the people are. I wouldn't encourage someone to do it who didn't know how, and I wouldn't discourage someone who did know how to do it. I think there is a danger here for people who are totally lacking in construction skills. They might blow it. That's the sort of person that I would discourage. On the other hand, there are a lot of people who are very clever even if they've never built a house before. They've put together cars or boats; they've got a good solid head on their shoulders, and they could do it. I wouldn't make any kind of policy statement except that if you're going to goof it up, you shouldn't do it."

"We emphasize in our projects, that a prospective home builder should not try and do structural work him- or herself but should hire out a contractor to do it because of the stresses involved," argues John Hand. "We do not sell plans. We retain a certain amount of control over the job. In other words, we like to sleep at night; we want to know what Joe Blow's doing out there. If one of the things collapsed and we had sold plans for three hundred dollars and they came back and sued us for four million, that doesn't work out too well. While anybody can come in and finish off the inside, as far as the concrete and steel work [is concerned], it's rather a specialized thing. Owners should turn that over to someone who's experienced with it."

John Barnard is more emphatic. "This is something you can't emphasize too

In the hands of a creative designer, the weight support requirements of an earth shelter can be translated into striking visual forms.

much. We get so many calls that 'Sonny and I are going to wing it, and we've built a chicken house once, and know how to mix concrete and wheel a barrow.' And this scares the hell out of me. We do find and we are getting guys who will get a competent contractor to put up the shell of it and reinforced walls, and get the top on it; then, depending on their degree of expertise, let them do the interior carpentry, the gardening, the painting, the plumbing, electrical wiring and heating if they know how. But there are darn few guys who can read a Popular Mechanics book and know how to place re-bars properly in a wall and how to anticipate all the openings that you have to have. And you can still save a lot of money by doing this properly."

If you're supposed to get together with a professional structural expert, how do you go about finding a good one? "That's a tough question," says Glenn Strand. "You can't direct them to one, you have to [make] allowance for people starting out in the business. At the same time, it's hard to beat experience. A building contractor or architect that has experience with the materials—there is a learning curve involved—will be more efficient. If you start with one from scratch, you're either not going to get your money's worth or else pay a bit to educate the person you're hiring to research it. In appropriate situations this happens all the time and works fine. Generally, people should look for someone with some experience and with some interest and enthusiasm. Enthusiasm carries a lot of weight. If there's one quality to back up experience, this is it. You get more work out of somebody who is enthusiastic. It probably ends up as much a gut feeling as anything else. You get more for your money if you hunt around and check all your options out. I think very few people are capable of designing conventional homes very well. Add earth sheltering on top of that and it's hard to beat experience and professional help. It could mean spending money to have him review your plans, or else just dumping the entire project in his lap and taking the finished product. But it's definitely worth it in the long run."

INSULATION

At the surface, insulation means wrapping the house as tightly and completely as possible. The air outside steals heat rapaciously. The sun pours heat into the house. The thin envelope of living space must be thoroughly separated from the environment.

Under an earth blanket, though, the situation is different. The earth environment, too, bleeds heat from the warmer house, but at a much slower pace than the air. In addition, the earth doesn't blow away with the heat like the wind; it stays put. Later, when the house cools, the soil can return some of that heat to the house. This gentle transfer is the key to earth-sheltered energy savings.

Therefore, earth shelters need to be insulated well enough to keep from losing significant amounts of heat to the soil in winter but loosely enough to allow for moderation of temperature with the surrounding earth. A number of methods have been devised to achieve this goal. All have one common trait never found on surface homes: They go on the outside of the building. Insulation on the inside would separate the interior from the concrete, which can absorb and store excess heat during warm periods and release it back into the house during cool ones. The outside insulation keeps the temperature in the home more moderate, with fewer fluctuations.

Insulation on the outside of a structure has to bear up under earth pressures and movement, roots, water and the rough-and-tumble period of backfilling. It can sometimes be protected by thin plywood sheets or felt, or even a longer-lasting nonrotting substance to make sure it doesn't shift during backfilling or give up the ghost to the elements later on.

Materials strong enough to withstand the outside environment cost more than fiber glass interior insulation, but they don't require any framing or carpentry for installation. So the cost difference is not great, but the performance edge is.

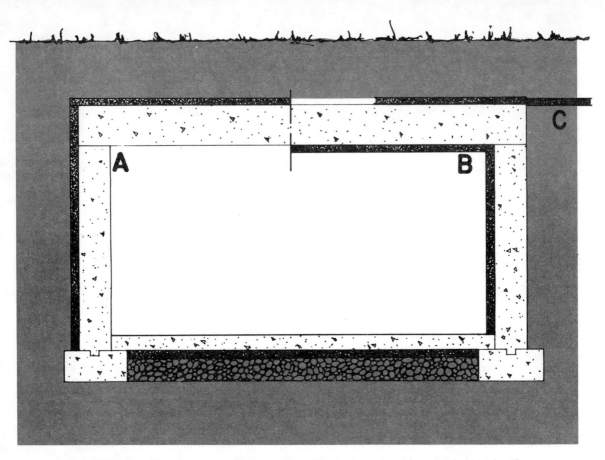

It is better to insulate earth shelters on the outside (A) than on the inside (B). Inside insulation lets soil moisture into the walls and traps dampness in the house. Exterior insulation retains heat without turning the interior into a steam bath. Insulating the top half of an earth shelter wall provides adequate protection in some cases. Extending the top insulation past the wall horizontally works best of all (C).

With that information in mind, here are some possible ways to insulate an earth shelter so it keeps its heat without losing touch with the surrounding soil:

- The American Underground Space Association found that beyond four inches of rooftop polystyrene insulation, additional insulation isn't cost-efficient. It saves some energy, but so little that it would take between eight and twenty-six years to pay back the cost of two extra inches of insulation.

- Because heat rises and the roof is the part of the house likely to be closest, and most exposed, to the surface, it requires the most insulation. The exact amount depends on the circumstances, but it's almost invariably more than the walls require, and since the roof doesn't have much of a relationship with the surrounding soil, insulation should reach from corner to corner.

- Some designers like the notion of wrapping the whole building, but they make the insulation steadily thinner as it goes down and the earth effect grows greater. Malcolm Wells advocates this approach, especially when the floor sits directly on the earth, because it keeps the floor warmer, reducing chilly foot troubles.

- Others propound heavily insulating the top half of the walls, leaving the bottom halves bare. However, Thomas Bligh found that a better arrangement is to extend the roof insulation beyond the walls, then drop a vertical insulation panel some distance from the wall, with soil in between. His tests revealed that a vertical insulation panel eight inches from the wall provided 109 kilowatt hours of energy savings during the winter and 100 kilowatt hours of greater cooling power in summer. His theory is that soil trapped by the insulation creates a protective cap over the building and increases its thermal mass, which moderates temperature change.

- A third scenario tested by Bligh extended the roof insulation about 5½ feet horizontally into the soil; there was no vertical insulation at all. Here, too, increased thermal mass is the object. And the scheme worked. The test case saved 54 kilowatt hours in winter heat and increased summer cooling by a whopping 233 kilowatt hours. It had the pleasant side effect of protecting the walls from surface water, thereby reducing the need for absolute waterproofing.

However, note that the last two insulation methods both involve construction problems and could become major design headaches where more than one wall contains windows.

The insulating material for any of these plans requires certain basic characteristics in order to perform properly in an exterior environment. The American Underground Space Association lists the following desirable elements for earth-sheltered heat savers:

1. Ability to withstand twenty to thirty pounds per square inch of earth pressure.
2. Great water resistance and low water absorption so the insulating value remains constant and unimpaired.
3. Toughness to stand up to soil acids and other chemicals for many years.
4. Ability to hold shape and insulating effectiveness for at least twenty years.
5. Tight-fitting, tongue-in-groove configuration so the panels lock together tightly, reducing heat leak and water seepage.
6. Low cost, availability and easy handling.

Two materials serve particularly well for earth-sheltering situations:

Extruded Polystyrene. This honeycombed rigid plastic repels water and stands up well to pressure and pounding. The most common brand is Styrofoam.

Urethane. If it doesn't get waterlogged, it can insulate a bit better—and a little cheaper—than polystyrene, but its structure isn't nearly as water-resistant. Also, some say that it may wear out before the requisite twenty years.

Because soils and climates differ so much, insulating techniques for earth shelters vary incredibly. Don Metz has pointed out why severe insulation is required even below grade, where he works in New Hampshire: "When the temperature gets down to 20° below, that earth freezes and it becomes a big ice cube," sucking heat from the house. It won't suck as much as the even more frigid surface air, but it'll suck plenty.

At the other extreme stands Frank Moreland, who recently built a home in Waxahachie, Texas, that will be under nine feet of soil and won't have a single centimeter of insulation because the ground will keep the temperature at between 63° and 69° all by itself at that depth.

And in places like Florida, where the important factor is how much heat the earth can draw from the home in summertime, insulation against the earth would be a detriment.

So keep in mind, when putting together plans for an earth-sheltered home, that there is no such thing as a standard situation. Earth shelter today still consists of interested individuals who are forging the standards for generations to come.

SITING

Where you put your earth shelter and how you locate it on your building site are considerations as important as the structure and design of the place. Since, by definition, earth-sheltered architecture requires greater interplay between the house and surroundings, the selection of those surroundings has to be handled with particular care. The relationship between a good underground design and its proper location is hand in glove, so any earth shelter—even a superbly designed one—on the wrong kind of lot will be like trying to fit a big hand into a tiny glove. Uncomfortable.

Earth shelters must suit their site. This home turns its protected back to winter winds, its face to the strong mountain sun, and uses the tree for summertime shade.

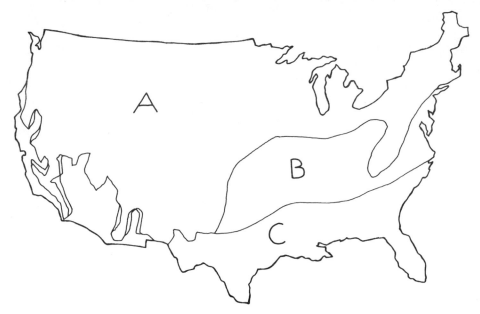

The map above outlines the three basic zones for earth sheltering in America. Section A offers, in general, the easiest climate for earth shelters. Seasons are fairly balanced and temperature variation is great enough to offer energy savings but not too great to create design problems. In zone B, earth shelter can provide both summer and winter benefits, but humidity might prove a problem. Zone C, the coastal temperature of the country, has the least need for earth shelters. The climate is mild all year, with high enough humidity to cause some problems. Often other design solutions will work as well here as earth shelter.

EARTH SHELTERING AROUND THE COUNTRY

First you must consider where in the country your subsurface building will be. Effective design varies with the climate, humidity and soil features of different regions. Earth shelters have been successfully built in every part of the United States, but the same kind of house won't necessarily work as well in Florida as in Minnesota, so you have to pay attention to the local particulars.

Here, from an architectural survey of America's earth-sheltered suitability, is an overview of the underground living situations in fifteen sections of the country, generally moving west to east:

Portland, Oregon: Basement floor temperatures are reported to range from 46° to 60°, requiring some heating (solar heating described as "adequate") for comfortable summer use. "Basements on slope-exposing wall on sunny side are suitable for living quarters." No specific reference to winter conditions.

Phoenix, Arizona: Atmospheric seasonal "design temperature" range is 16°–106°, pointing up the desirability of ameliorating devices. "In this region basement might prove to be most comfortable living portion of house. Several feet below surface mean annual temperature of 70° is present both day and night in winter and summer. This is an ideal living temperature, and by building down into the ground this temperature should prove to be an asset in maintaining constant living conditions."

Denver, Colorado: Basement described as "desirable," that is, cool in summer (no humidity problem) and easily heated in winter. Optimum condition would include sloping site with southern elevation fully exposed to the sun.

Twin Cities, Minnesota: Atmospheric design temperature reported to be $-12°$, compared to minimum basement design temperature of $+31°$. Southerly-exposed wall is recommended, largely in response to the relatively high humidity of the area. Bligh uses Minneapolis region as example for demonstrating energy-conserving benefits of underground space use.

Mid-Mississippi Basin (St. Louis–Kansas City): Subsurface rooms, with ground temperature constantly at $55°$, will conserve considerable fuel in winter because floor slab will always be $20°$–$40°$ warmer than outside air. Subsurface rooms, if properly dehumidified and ventilated, will be most comfortable part of house during hot months.

Chicago, Illinois: "If properly dehumidified, basement rooms will be attractive retreat during summer months." Severe Chicago cold and wind not discussed, but points up obvious benefits of winter use.

Columbus, Ohio: Minimum basement design temperature given as $30°$ versus $8°$ for outside air, thus providing relative winter warmth. Humidity is indicated as a concern, but "if humidity is controlled, basement will be most comfortable part of house during summer months."

Pittsburgh, Pennsylvania: "Basement living quarters would cut fuel requirements for degree days approximately in half in winter, but would require additional vapor sealing and air circulation." "Normal summer temperature in basement too low for comfortable living conditions. Some summer heating required, and a southerly elevation exposed to sun is suggested where possible."

Boston, Massachusetts: Humidity and condensation problems are cited, which may be at least partially alleviated by some sort of solar heating (e.g., southern wall exposure or extension of wall to provide light penetration into window walls). Ventilation alone inadequate check on dampness; no other notes, although severe winters suggest considerable heat-conserving qualities.

Albany, (Buffalo–Montreal), New York State: "Potentially a basement in this area has superior advantages for living facilities, for which it is cooler in summer and warmer in winter, and if these lithosphere rooms were made attractive and spacious, they would probably be preferable to living quarters normally planned for floors above the ground." Solar heating—by means of light wells and conservatories—is suggested both for winter warmth and as a means of dealing with occasional high summer humidity.

New York Metropolitan area (including Philadelphia): (Conditions similar to Columbus, Ohio, area.) A basement design temperature of $30°$ for winter is contrasted with atmospheric design temperature of $12°$. It is stated that "unheated basements will be warmer and drier than the outside atmosphere," 47° and less than 65% relative humidity being present during the winter months.

Washington, D.C.–Chesapeake Bay areas: Basement areas are said to provide maximum summer comfort (if properly dehumidified) and a relative source of heat in the winter. Solar heating methods are recommended, particularly to deal with summer humidity. (Ventilation may only increase it.)

Charleston, South Carolina: High humidity and relatively high summer ground temperatures make basement living areas unsuitable for use during the predominantly warm seasons; no notes on winter benefits, but the generally moderate climate makes few rigorous demands on the building as a whole (stilts suggested).

Gulf Coast (Florida–Texas): Basements are generally omitted. "High humidity, combined with high ground temperatures (about 70° in summer) make underground areas unusable for living or storage." Instead, "the higher the living quarters are placed, the more comfortable they are likely to be."

Miami, Southern Florida: High groundwater and ventilation requirements exclude basements from consideration: "A basement would be a liability because of high humidity during most of the year."

In general, warmer regions have gotten something of a bad rap as far as earth sheltering goes. But, as Don Metz notes, going under the soil is not merely a cold weather phenomenon: "Up north the heating system may have to make up just a 20° difference, but in the south earth sheltering actually provides a temperature differential in your favor; there's nothing to be made up at all. For instance, when the air temperature down south is 100° in summertime, the temperature down in the earth is just 70°; the earth heat sink is pulling 30° out of the house without using any air conditioning. But in wintertime up north the earth might be 45° when we want the house to be 65°, so we have to make up the difference. In that sense, earth sheltering makes a better bargain in hotter climates. And, in fact, all the hot, dry countries have an indigenous tradition of earth-sheltered buildings. Tunisia, China, North African countries, Mediterranean countries—all have earth-sheltered cave towns. The original troglodyte, I'm sure, was Mediterranean."

DESIGNING FOR THE ENVIRONMENT

A region's environment might present problems in designing an earth shelter, but few situations absolutely preclude living with the soil. If the structure is created to blend with its surroundings, earth shelter is possible from southern beach to northern mountaintop.

A successful seaside design, such as the one shown in the accompanying illustration, counters the drawbacks of a beach environment with natural ventilation, a shell-like contour to spread the weight of heavy, shifting sand evenly, deep shading, and infiltration of moisture through the house to maintain groundwater levels.

In temperate, flat areas, a southern-oriented two-story elevational bermed design offers maximum heat-saving potential, a full drainage system for spring rains, cross-ventilation for summertime cooling, and insulation levels that permit a steady heat flow with the surrounding soil for balanced temperature. Such a design is shown in the illustration on page 78.

In a hot and arid climate, a sunken, open design with an earth-covered roof provides strong ventilation to keep things comfortable during the scorching summer, as well as large awnings to shade the house from the summer sun (see illustration). Water pools trap heat during the day and radiate it at night, keeping the house cooler during the daytime, warmer at night. During the mild winters these pools can store some heat to warm the house at night. By being situated slightly lower, the house avoids the strong sun, gets better surrounding earth-temperature patterns, and escapes the worst of the desert winds.

A mountainside earth shelter might be set back into the slope, as in the design shown in the illustration on page 79. Heavy drainage and insulation on the exposed roof reduce water and heat problems. Ventilation is created by an open-close skylight that draws air upward during summer, bringing cooler lower air in behind. Solar collectors on the roof heat water and take the edge off home heating during wintertime; rock storage for solar heat can be placed beneath the floor, serving a double purpose of warming the floor directly and warming the air to help heat the rest of the house.

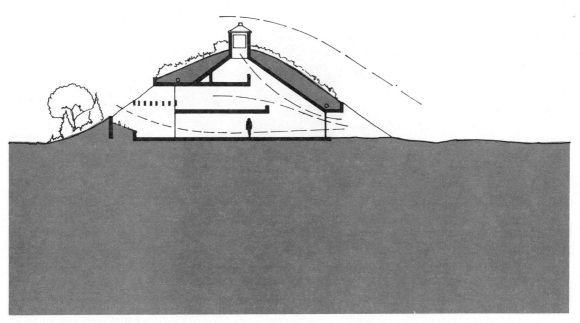

A successful design for a hot, humid oceanfront climate should feature extensive natural ventilation, excellent drainage, deep shade and a site orientation that makes the best possible use of the limited land contour.

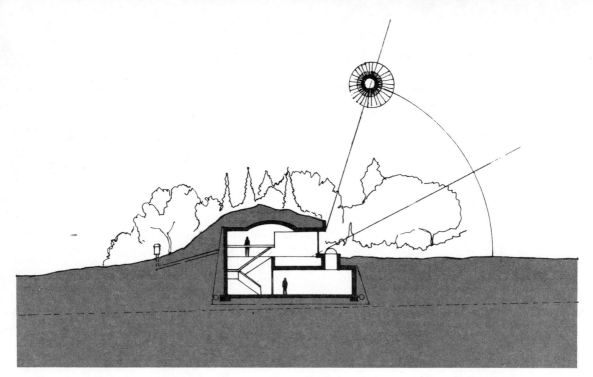

A temperate zone design must pay attention to both underheated and over-heated periods. The balance can be achieved by flexible solar heating capabilities, radiation loss and convection into the earth. This house features maximum ability to gather solar heat with maximum ability to share temperature with the earth.

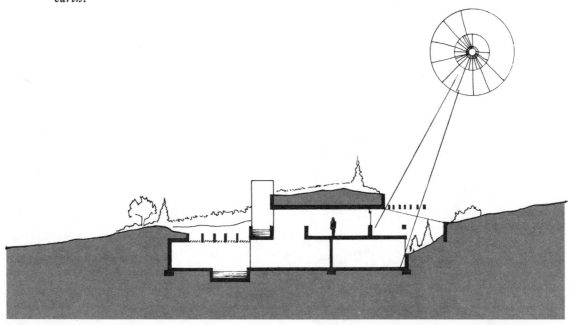

Where the environment is hot and flat, the house must fight off extreme daytime heat. This design is set deep into the earth for greatest heat moderation. It has heavy earth cover on the roof to enhance the effect. At the left side is a water pool. In summer the screen above it can be closed during the day and opened at night so the pool can radiate heat from the house to the cool night sky. In winter the pool can be open to the sun during the day and radiate the accumulated heat into the house at night.

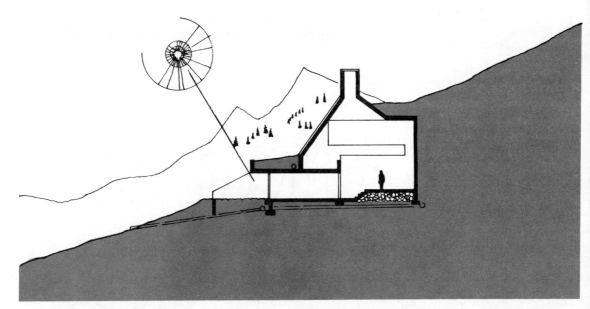

In a cool, steeply sloped region, drainage is a major concern. The house should be situated where runoff and erosion can be controlled. Extensive solar exposure is needed during the winter and high ventilation during the summer. This house is heavily insulated to retain heat in winter, but it is capable of being opened up in summer for great natural ventilation.

SUN AND WIND

With energy savings as the goal of earth shelter, and passive solar gain an important step toward that end, the proper north-south orientation of the house is also important, whatever region it stands in. An earth shelter can exist on a different orientation, but a south-facing design will always be more efficient. This is so because in the United States the sun rests in the southern sky. An earth shelter with its face turned toward this free heat source and its back protected against north winds obviously nets a heating boon. A design angled to east or west either loses the benefit of the southern sun—and thus costs more to heat—or must be cleverly constructed so that the roof openings face the sun even if the house doesn't. Malcolm Wells's house (described later in this book) is an example of how this can be done. More on the relationship between earth building and the sun can be found in the next section of this chapter.

Light and heat are not the only important considerations in orienting an earth shelter. A house should be situated so it has access to summer breezes, which generally come from the southeast, and is buffered from the bitter winter winds that whistle down from the northwest. (Specific wind features vary from community to community, of course; these are the general trends.) Does your site have what Malcolm Wells calls "a blanket of earth over its northwest shoulder" to divert those winter winds? Does it have an open access for southeasterly summer breezes? Do the natural formations form a wind tunnel that might subject you to howling gales at the slightest provocation? Are you blocked in so that you'll stifle with heat and humidity during the dog days?

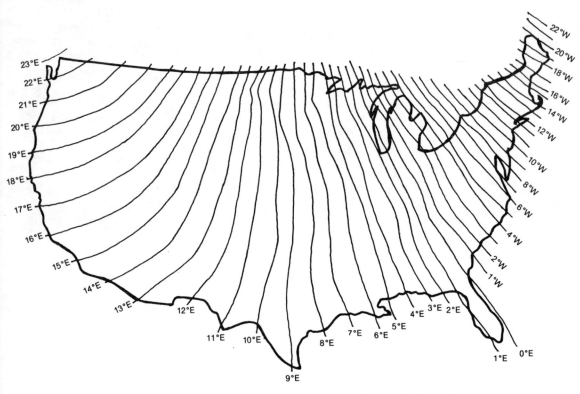

Southern exposure is important for many earth shelter designs, but a magnetic compass doesn't usually point to true north. These lines show how far a compass will vary from true north in different parts of the country. Adjust your design accordingly.

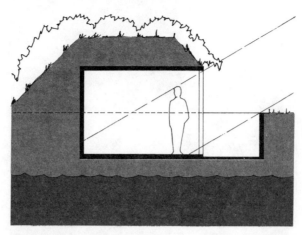

The first illustration shows the house set so shallowly that there isn't enough fill to cover it properly.

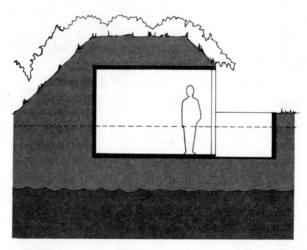

In the second illustration the living space is in much better balance with the surroundings: light reaches well back inside and the water table lies well below.

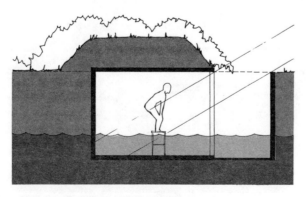

The third drawing has the house set down too far. Seeing outside is hard, light doesn't reach all the areas, and water will be a problem.

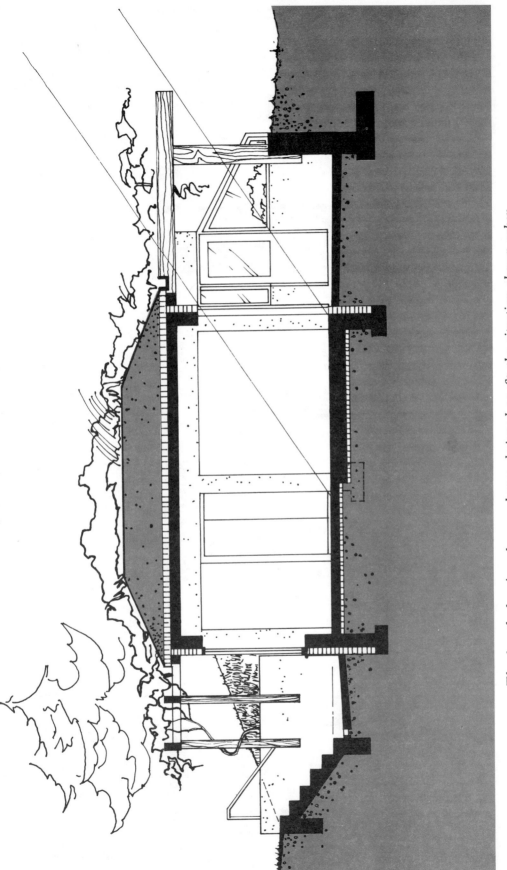

The fourth drawing shows a house designed to fit the situation shown when elements are in balance.

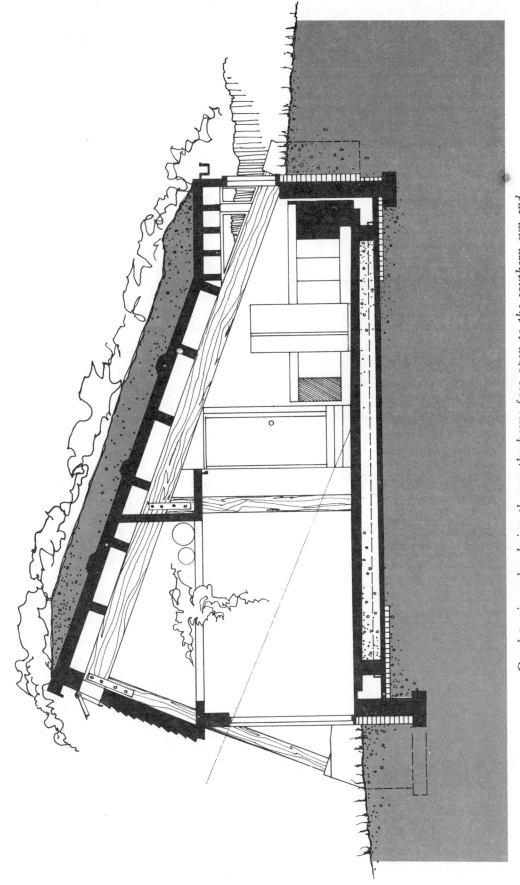

Good passive solar design throws the house face open to the southern sun and turns a protected, stingy shoulder to chilling north winds.

VIEWS

Next, what about the view your site offers? First look to the south. If the view lies in that direction, you've got a peach. If, however, the best view is east or west, you have to decide if seeing it is worth giving up a good deal of energy savings by adding a second wall of windows. By the same token, appraise the rotten views from the site—a factory, highway or rail line, for instance. Can the house be situated on the site so it looks away from these eyesores?

The flip side of your view is how much or little your neighbors can see of you. If your huge southern exposed wall stares right into the side of a neighbor's east-facing house, he won't be happy and neither will you. Can surrounding houses on higher ground see down into your window wells or skylights? Certainly you don't want to feel on display in your home, so check this situation out thoroughly before deciding if a particular lot is for you.

OTHER FACTORS

In addition to what has been noted thus far, every factor discussed in this chapter impinges upon site selection. You should be familiar with all the important elements—soil, water, structure, and so forth—and consider how these pieces fit into the puzzle of site selection. Other items to keep in mind include:

Animals. If you're building in the country, remember you'll be close to, or in, the ground. You won't want bears sitting on the skylight or snakes trying to worm their way into the house. Find out about local conditions before it's too late.

Vegetation. We've already mentioned evergreens that can block summer sun. Pines that stand near the northwest end of the house can block winter winds and should be much prized. A stand of deciduous trees to the south of the house can bring you shade in summer but let the winter sun through the branches. Also, does the surrounding vegetation indicate that your ground cover and garden will flourish?

Utilities. The site may look lovely, but where are the power lines? the telephone poles? Are there any underground cables that will interfere with your excavation?

Sewage. Are there sewer hookups close at hand? Will they run above the floor level of your house, making sewage elimination a problem? If you are going to go the septic system route, is there a good spot around for the leaching field?

Site Access. Will you need to build an access road or highway? How practicable will that be? Will it cost a lot? Where will you park your car? Do you want a surface garage? If so, where will you put it? Lastly, will construction equipment necessary to build your house be able to get to the site? No sense buying it if you won't be able to get the builder there.

One of the less frequent criticisms of earth shelter design is that the process of construction scars the building site more than that of surface homes. All that digging. All that concrete. It seems like such a shame to dig up a great plot of land. Malcolm Wells, for one, agrees but feels that this is no impediment to earth sheltering if you pick a site not for its looks but for its view. "The best site, to my mind, is useless land. We can put underground houses on land we don't want for anything else. We can leave the good land for looking at and playing on, which is how it should be."

EARTH SHELTERS AND SOLAR HEAT

When Paul Shippee, of Colorado SunWorks in Boulder, wanted to design an ultimate solar passive home, he selected earth sheltering as the most natural way to go. In the spring of 1978 SunWorks unveiled an 1,820-square-foot home in Longmont, Colorado, that combined earth shelter and passive solar ideas into a striking vision of the energy-niggling future.

The home has eight-inch-thick concrete walls, with three hundred square feet of double-pane glass on the southern exposure. The center of this huge expanse of glass lets sunlight into the house; the rest of the exposed wall is covered by water-filled fifty-five-gallon drums painted black. During the day, as the sun beats down on the glass, the water in the heat-absorbing black drums grows warm. At night, as the windows cool, insulating Styrofoam beads are blown into the space between the window panes, holding the accumulated heat inside. The drums radiate their heat into the house at night, keeping the place cozy.

Computer surveys of the SunWorks house have shown enormous energy savings. "Throughout our extremely cold 1978–79 winter, the backup furnace was never used," Shippee states. During the winter of 1979, waste heat from appliances, lights and residents provided 17.2% of the home's heat, while the passive solar arrangement netted the rest. The temperature inside the house ranged from 63°—on a day when the outside temperature plunged to −20°—to 81°.

In summertime the SunWorks house uses angled shades to block sunlight from the southern wall and natural nighttime ventilation and heat radiation to keep it comfortable.

This demonstration project shows how solar heat and earth-sheltered design are natural partners. The strengths of earth sheltering shore up the weak spots in solar heating systems and vice versa, producing a combination that performs better than either unit separately.

In aboveground structures the gentleness of solar heat can be a problem. The sun's diffuse radiation may be able to generate heat, but not always enough to warm a large surface home exposed to heat-stealing winds and weather. In wintertime the differential between the surface temperature and indoor comfort can be more than 80°; solar heating systems can't often generate that much thermodynamic clout. So the percentage of heat they can offer is low.

In an earth-sheltered building, however, the total amount of heat needed is far less, so the percentage of heat that can come from a solar heating system is far larger. As the next section on Energy Savings notes, in an earth-sheltered dwelling the range of temperatures from season to season will seldom vary more than 15° or 20° all by itself in midwinter, but it surely can raise the temperature from 50° without much trouble.

The moderating heat sink of earth, which is the key to earth shelters' energy efficiency, also increases the effectiveness of all types of solar heating systems.

Another happy coincidence between earth sheltering and passive solar systems in particular is the enormous thermal mass of underground buildings. In the section on Structure the need for strong, heavy concrete or block walls and floors was explained in detail. Earth-sheltered homes must be strong, thick and carefully insulated and water-sealed in order to coexist with the soil. These requirements not only make the house strong, they stabilize the temperature as well.

In a surface house, with its thin walls and constant exchange of heated air with the chilly outdoors, temperature changes happen fast. When the sun beats on the roof, the house gets very hot and must be cooled. When the wind steals heat, it races away and the temperature must be constantly shored up. A surface home has little capacity to absorb and store heat.

An earth-sheltered home, on the other hand, with its thick walls, absorbs heat slowly. On a hot day the heat from passive solar windows or skylights raises the temperature of the building very slowly. Instead of creating a boiling house at noon, as would happen in a surface house, the sun will produce peak temperatures in the house only toward the end of the day. This energy, stored in the walls, floor and ceiling, slowly radiates back into the house all night long; it doesn't seep out quickly, as on the surface. So an earth-sheltered building puts whatever heat the sun provides to better use than a surface home. And it needs less heat all around to remain comfortable.

Such energy relationships should become increasingly important over the next few years as federal Department of Energy Building Energy Performance Standards (BEPS) come into play. By 1984 the agency proposes, BEPS limits should be enforced on all home construction. Many architects feel that the likely ceilings for BEPS performance will be stringent enough to force rapid development of all types of homes that work well with passive solar designs. Earth-sheltered homes will present just about the best possible set of performance criteria among passive solar possibilities.

Of course, passive solar heating systems are not the only ones that work with

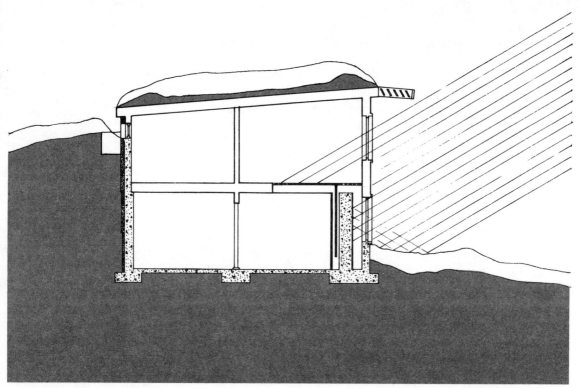

An award-winning passive earth shelter design. Two stories are buffered by earth on three sides; both are open to the sun in the south. Warmth is captured by the second-story floor and by a thick trombe wall on the first floor, which captures hot air between the windows and the room until it gets hot enough to rise over the top into the cozy house.

earth-sheltered homes. All typical active solar heating systems are compatible with soil-buffered buildings, too. Solar collectors can be placed on the south-facing sloped roof of a sunken home, just as they can on a surface one. Here, too, they have to produce less heat to make the home comfortable than they would on the surface.

Active solar heating systems work excellently at providing hot water for earth-sheltered homes. "Using active solar [heating] systems for underground domestic hot water is cost-effective at this point," states engineer Thomas Bligh, the leading structural expert on underground architecture. "You can use tax breaks and do some of the installation work yourself and pay back the cost of all the solar equipment in five to seven years easily. That's cost-effective, I would say. I've even seen some domestic hot water systems that have paid themselves back in two or three years, which strikes me as being darn good."

Using active solar heating systems for earth-sheltered houses strikes many architects as being too expensive and elaborate a solution to a simple problem. "I tell customers," says John Barnard, "look, you start in with a low energy consumption on a sunken house anyway. When you cut it in half, you really don't save a hell of a lot. Now, any active system that I've seen will cost at least ten thousand dollars for the hardware and storage material and all the extras you need. You end up spending ten thousand dollars to save very little. For underground design passive solar makes a lot more sense. It does what needs to be done and it can be designed into the structure from the outset at very little extra cost."

Malcolm Wells agrees. "Our preference is for the simpler approach. We prefer to work with systems that are easy to understand, systems we can fix with a few tools and some materials from the local hardware store. If we were scientists or engineers, we might feel differently about this. Notice we said 'might.' More and more of the scientists we respect are moving toward simplicity, and since that movement is so compatible with underground architecture, we have a double reason for endorsing the uncomplicated approach."

Not only does earth sheltering increase passive solar's effectiveness by reducing heating requirements, it situates a house in such a way that all the best parameters for staving off cold winter winds are achieved. As Malcolm Wells notes, "Sometimes it is simply hard to believe that so many good things could happen when we pull the great earth blanket up over our north shoulder and welcome the sun at our south side."

The hill behind a southern-facing earth-sheltered home may deflect bitter winter winds off into the night, but the effectiveness of any passive system can only be fully maintained through savvy insulating. Insulation of earth-sheltered homes has its own section in this book, but a few insulating tricks go particularly well with passive solar.

Insulated shutters that fold up during the day and slide across windows at night can make sure the precious heat garnered during sunny periods stays inside the house, where it belongs. On the flip side, the shutters can keep out the hot sun of summertime so the cooling earth can keep the house comfortable.

Keeping windows on the sides of the house small is another good adjunct to a passive system. Obviously, the best energy savings of all comes from having no windows at all, except on the south wall, but that is claustrophobic. Better to let in some light and lots of view in all parts of the house and give up a tiny bit of energy. These windows, too, can be shuttered at night. But during the day they allow you to look out in a direction that doesn't face directly into the blinding sun, which makes earth-sheltered passive life much more pleasant.

Another passive possibility is the Trombe wall, which works well in earth-sheltered contexts. This is a large masonry wall painted black that stands directly between the southern exposure of solar glass and the home's living area. During the day the wall gets hotter and hotter. At night the window is shuttered and vents in the wall are opened. Air flows over the wall, is warmed and then heads into the house. The wall is basically a giant solid heat-storage tank that saves up sunlight for use at night.

The most intriguing passive solar idea I've heard, one that is specifically geared for earth sheltering, is earth pipes. These consist of an array of heat-conducting pipes that are laid beneath the floor of an earth-covered house. They are connected to a duct in the roof by a fan. During the summer the fan draws hot air off the roof and down through the pipes. Back and forth goes the hot air, slowly giving up its warmth to the ground. All summer long, heat that the sun dumps on the roof is transferred by the pipes into the ground under the house.

The air is eventually blown out a duct at the back of the house, but the heat remains in the soil all summer long. Come winter and the house stands atop a preheated patch of earth that radiates summertime heat up against the floor. Now, heat gained by earth pipes certainly won't heat any home—even an underground one—by itself. But it could mean the difference of a few percentage points in energy consumption at virtually no cost.

That's what solar heating is all about.

Maximum exposure to light and sun in a large earth shelter home. Trees screen the house from the north winds. Earth and vegetation above the garage protect it still further. All windows and skylights angle toward the southern sun.

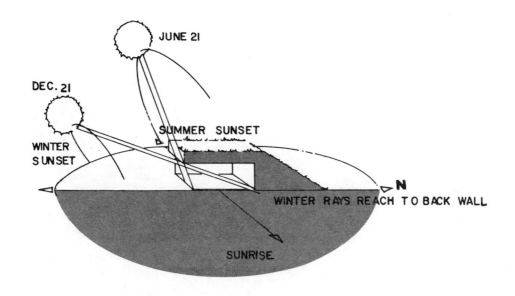

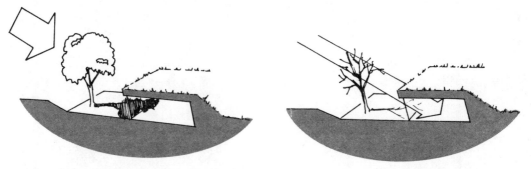

The house's orientation and passive solar. At top the house is designed facing south, with an overhang that shades the higher summer sun but lets in the lower, necessary winter sun. At bottom a tree shades the home's southern wall in summer but lets the sun filter through leafless branches in wintertime.

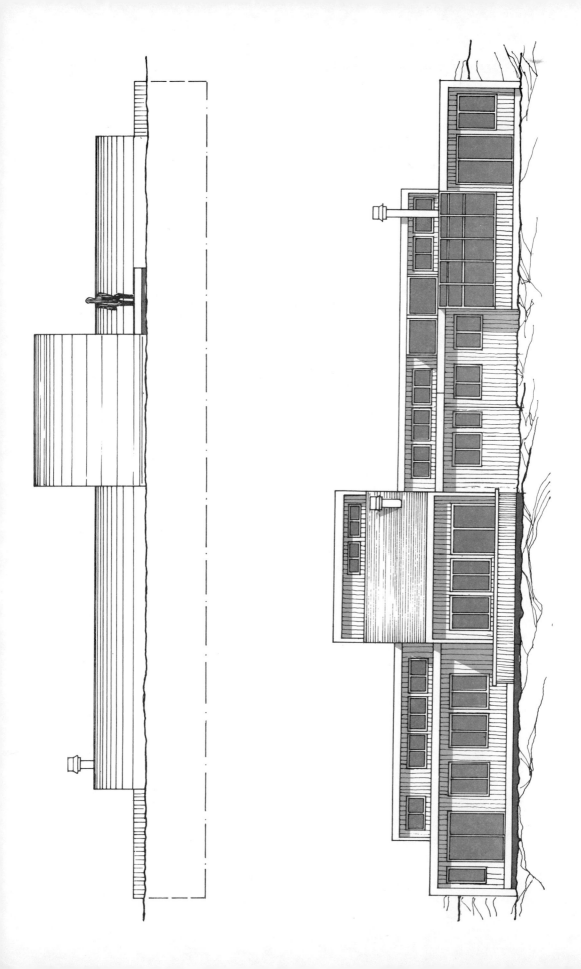

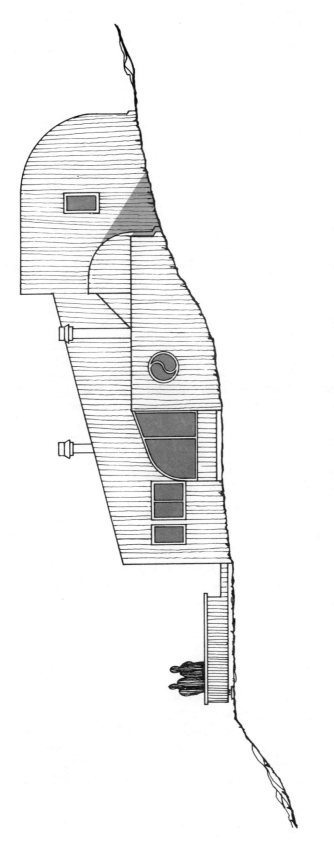

Three views of a passive solar earth shelter. The north side (top) is nearly non-existent. Winds have little to buffet. The south side (middle) is completely exposed and full of windows. The east and west sides (bottom) have small windows toward the south and are tucked into the hillside as far as possible.

ENERGY SAVINGS

As insulation earth stinks. Foam insulation holds heat back some twenty times as well. So why does an earth-sheltered house use less energy for heating than a surface home cozied up with insulation?

For the simple reason that while the earth is lousy for insulation, it's great for retaining and moderating heat. It takes a long time to heat up soil, but once the earth absorbs heat it holds on to it for a long, long time. The temperature of the surface air jumps up and down at the drop of a cold front, but the soil temperature slowly rises and falls in a steady, mild, predictable fashion.

The air that races around a surface dwelling is thin and impermanent; yesterday's breeze from Cleveland is winging its way past Akron today, headed for Pennsylvania tomorrow. It carries with it heat stolen from every building in its path. The soil around an earth shelter, however, stays put. Heat that leaks into it from the house stays near the house. Heat placed in it by the summer sun stays in it well into the wintertime.

THERMAL MASS

Soil has great thermal mass. This means that soil can absorb an enormous amount of heat during the warm season and, because of its bulk, can hold on to that heat for a good while despite cold temperatures around it. Think about it: In winter, frost and a layer of snow insulate the lower soil; whatever heat is down there stays down there because it can't get to the surface easily at all.

The greater the thermal mass, the more moderate temperatures in an environment will be. Mass wipes out big swings in temperature. Concrete and stone

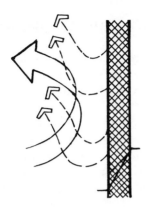
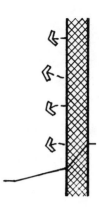
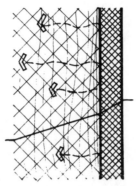

The comparative heat effects of wind and soil. Above left, moving winds draw heat from the wall of a house. Above right, motionless air currents don't steal heat quite as fast. The least amount of heat escapes when there is a surrounding layer of soil, shown in the center illustration. The solid lines that angle through the illustrations show how fast the inside loses heat to the outside in each situation.

have more mass than wood. Water, oddly, has greater thermal mass than anything. That's why water-filled drums are so popular in passive solar designs; the water accepts heat better and releases it more evenly than any other natural substance. But floating your house in a water bath seems slightly impractical, so earth offers a reasonable alternative.

What is so important about having thermal mass gobble up and store heat? Well, do you recall those power company ads that bemoan the necessity of maintaining generating capacity for the rare peak summer and winter demands? If only we could use more power during off-peak times, they say, we could save great sums on power plants and lower your electrical rates.

Heating works similarly. A surface house has to have a heating system ready to lift the temperature inside all the way from frigid outside levels to 68° or so. In summer the cooling system has to take the edge off perhaps a 100° interior swelter. An earth shelter of similar size in a similar area will be able to use a far smaller heating and cooling plant because thermal mass will keep the temperature variation between seasons much lower than on the surface. In addition, since the plant has less temperature to produce, it can work less often. And the ground conserves energy that gets out of the house, so the heating system becomes still more efficient.

These are the reasons for the significant energy reductions in earth shelters.

A good specific example of this comes from Ray Sterling, of the American Underground Space Association: "In Minnesota," he notes, "we have an air temperature range of 130° F annually: from −30° in winter to 100° in summer. But the temperature of the soil, if you dig down just ten feet, swings only 20°—from 40° to 60°. Even immediately below the surface you have only a 40° annual swing. And there's practically no daily swing."

Another startling example comes from John Barnard: "A couple of Februaries ago," he recalls, "the owner of one of my houses in Wyoming was having some trouble with his furnace, so before he went away for a week he

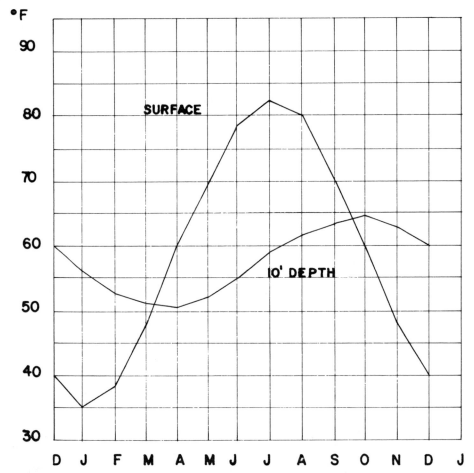

Earth shelter savings in Lexington, Kentucky. This graph compares the temperature at the surface with that ten feet down. It shows how much less temperature variation an earth shelter must contend with. (The letters at the bottom of the graph represent the months of the year, starting with December.)

just turned the thing off. Outside temperatures went well below zero for several stretches. But the house never got below 55°." Now that's temperature moderation.

And James Scalise weighs in with the most astonishing example of all: "In a study of earth temperatures taken over a ten-day period of sunny days in August in the Gobi Desert, the average temperature for a twenty-four-hour period, .08 inches below the surface, ranged from 36.7° F to 101°. At just twenty inches below the surface the temperature ranged from 52.22 degrees to 53.7°."

In other words, the average temperature at the surface usually comes from balancing two extremes—summer highs and winter lows. The average beneath the surface may be much the same as that up above, but it comes from a series of days that all lie right along the average. In Wampum, Pennsylvania, for instance, comparison of an 18,000-square-foot underground research lab with the surface environment shows the difference well. The surface average was 59°, based upon a summer average of 79° and a winter average of 38°—a spread of 41°. In the earth, on the other hand, the average was 54°, based upon four seasons of days hovering right around that temperature, with the result that the underground structure took half as much energy to heat.

Not only is the range of temperature better below grade, but the temperature pattern facilitates comfortable living as well. This comes from the so-called thermal flywheel effect. The thermal flywheel is a byproduct of the thermal mass of the earth. The mass of a physical flywheel keeps it spinning long after a lighter wheel would have stopped; similarly, the mass of the earth keeps it warm long after the air has turned cold.

This means that the peak and low temperature points for soil don't coincide with those of the air. While the surface hits its coldest moment in January or February, the earth doesn't cool off until April. And while the surface swelters in July, the soil is still ambling toward its heat peak in September or October. Thus, in summer the soil is cooler than the surface and in winter it's warmer. In summer the cool soil can draw excess heat from the house and in winter (depending upon location) it can either feed heat back into the house or at least take the edge off heating needs.

The heft of the thermal flywheel grows with thermal mass, and thermal mass grows with the amount of soil surrounding the house. Gordon Moore, an earth shelter engineer at the University of Missouri, has found that the temperature extremes in the soil shift seven days from the surface norms for every foot downward you go. "The goal," according to Moore, "would be to get completely out of phase with the season. You could get heat from the soil in winter and cooling in summer if you went deep enough." Even if you don't go deep at all, the thermal flywheel effect will save energy in an earth-sheltered structure.

SAVINGS

How much energy do you actually save? Thomas Bligh performed the classic study: "In our study we looked at different kinds of houses that have been monitored rather carefully and then we projected for an underground house based on computer simulation. These houses all used conventional construction, were all the same size, and were all adjusted for the same number of degree days —based upon Minneapolis. If all these houses were built in Minneapolis, this would be the result: The Twin Rivers house, which is a conventional house, used 55,000 kilowatt hours in a year. We then looked at a conventional controlled house built in Maryland. That house was about the best you could do with conventional construction: good quality, double-glazed, well-sealed conventional windows, vapor barriers, slightly more insulation in the ceiling. And it used 35,000 kilowatt hours. Fifty-five thousand down to 35,000. We then looked at an energy-efficient house in Maryland. It could be called a super-insulated house, with R19 insulation in the walls and R38 in the ceilings, super vapor barriers, fewer windows on the north side and more on the south, a heat pump. (R is a technical term for relative resistance to transmitting heat.) Here energy use went down to 21,000 kilowatt hours.

"Then we looked at an underground house. And we used exactly the same amount of energy for lights, domestic hot water, refrigerator, stove, dryer, et cetera, as the other houses did. In fact, we matched it to the energy-efficient house. And instead of using 21,000, the earth shelter used 11,500.

"On a percentage basis, if the regular house equals 100%, the good, conventional structure was 65%, the super-insulated house was 39% and the underground was 21%. And if you put thermal shutters on the windows, this dropped down to 18%. The total amount of heat used for this house was less than the heat used in domestic hot water above grade."

In another survey Bligh found that in 25°-below-zero weather heat loss through a surface wall with eight inches of insulation was six and one half times greater than through a noninsulated earth-sheltered wall. "An underground wall will be better than the best aboveground wall every time," he says. "In no way can improved insulation on an aboveground building begin to compete with subsurface structures from the viewpoint of energy conservation."

John Barnard feels that 60% reductions in heating requirements are easily achieved. And James Scalise states that just a thin layer of soil on a house's roof in conjunction with berming up to the windows will reduce energy consumption by 30%.

It should be clear, though, that experience with earth shelter is new enough that energy-saving pronouncements are still tentative. Bligh's are based upon surveys and those of the architects are based upon limited experience. This doesn't mean they are wrong, just that they aren't as solidly grounded as they will be after a few years of field experience. The field is just now getting its first set of genuine national standards for measuring component performance, so figures will shortly become far more reliable than the hip-shot predictions of the past.

Also, as with EPA car-mileage ratings, your energy savings underground will vary with your habits. An architect can't guarantee exact figures from construction alone. "All we have are consultants, mechanical engineers, solar engineers. And they give us their best estimate of what kind of energy savings you

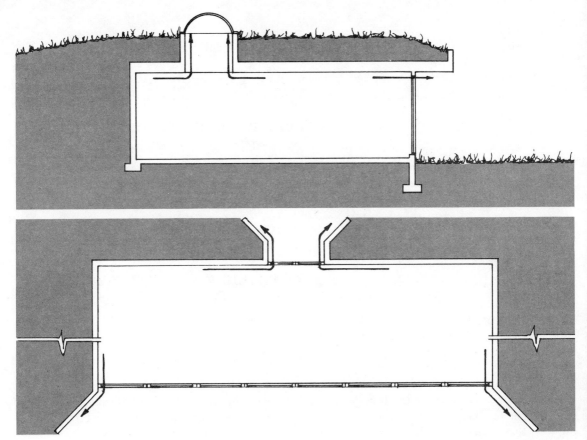

One way to maximize earth shelter energy savings is through the use of thermal breaks. These consist of layers of materials that don't conduct heat placed at spots where heat is most likely to sneak out. This illustration shows the escape routes of heat in a typical earth shelter and where the breaks should be placed to stop it.

get," says John Hand. "They can better predict heating and cooling needs than they can how many people are going to use an electric toothbrush. But generally we are looking for reduction in the 60%–90% range. According to the size, the kind of structure, the type of earth cover, the site it's located on, it would be possible if people were willing to wear sweaters some times of the year and T-shirts in others to stretch their temperature-comfort range a little bit. Then they might never turn any heat on. But we don't have any data to say that in general you don't need any backup thermal system."

Here is a way for you to approximate the savings you might receive from earth sheltering, compared to surface living, based upon a study by Lloyd Harrison. He compared a fifteen-hundred-square-foot surface house with an earth-sheltered one-story design of the same dimensions. The climate was assumed to be similar to that of Denver, Colorado. Without doubt the climatic factors elsewhere in the country would change these figures somewhat, but since Denver has fairly average weather, they provide a reasonable guide to work with.

Harrison found that the underground dwelling generated a 72% energy saving for heat. Specifically, he found the following rates of heat loss and gain:

	SURFACE HOUSE	EARTH SHELTER
Heat loss in winter		
(British thermal units per hour)	39927	12720
Heat gain in summer		
(British thermal units per hour)	44650	0

Translating the demands that result from these figures, Harrison determined that the houses would require these amounts of various heat resources:

	SURFACE HOUSE	EARTH SHELTER
Gas heat (cubic feet)	93828	30777
Oil heat (gallons)	710	233
Electrical heat (kilowatt hours)	23157	7596
Electrical cooling	3962	0

To get an idea of the amount of money an earth shelter might save you, get the current local rates for the energy system you plan to use. Multiply the amount shown above by the rate. The difference between the surface and earth shelter amounts will be a general indication of your savings.

HEATING SYSTEMS

Because of the lack of official data on the heat capacity of various earth-sheltered designs over a period of time, it may be necessary to engineer your heating system larger than necessary—better safe and a bit more expensive than to cut corners and shiver all winter. However, once the official standards are in place—a matter of months, not years—architects all over the country will have accurate readings to figure from, so they will be able to make your heating system just the right size for your situation.

As for that heating system, several different types of heat generation can be used:

Nothing. In warm areas wintertime heating may become totally unnecessary in an earth shelter. The ambient temperature of the ground, plus body and waste heat, may be all that's required for comfort.

Passive Solar. Read the section on Earth Shelters and Solar Heat for details. By designing with the sun in mind, enormous heat benefits are possible below grade.

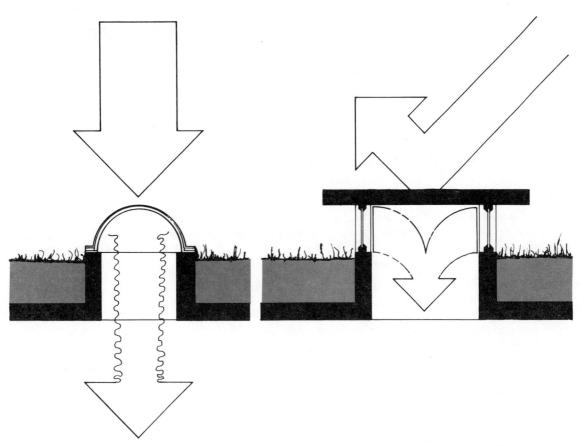

Two methods of solar heat gain. At left a skylight produces both trapped solar heat and direct sunlight for the interior. If more diffuse light is desired, a system such as the one at right can be used. The absorptive cap radiates heat into the house but blocks direct light. Gentler light enters through the sides.

Wood. In rural regions where wood is plentiful and reasonably priced, this offers an attractive possibility. A wood stove can heat a whole earth shelter even in bitter winter climates because the amount of actual heat needed by the building is small. The drawback with wood stoves on the surface is that when they begin to die down during the night the temperature in the house plummets; that won't happen under an earth blanket. Wood, on the other hand, presents draft and ventilation concerns. Such a system won't help with summer cooling at all.

Heat Pump. This is the most efficient heat engine yet developed. It uses a compression principle to literally squeeze heat out of even the most frigid air, which it then releases to house air at a higher temperature. The system is electrical, with no other fuel required at all. Architects greatly favor the combination of a heat pump with forced-air distribution for earth shelters.

Oil or Gas. You can use these, of course, but there are problems. First, your earth shelter will be more enclosed than a surface home, so leaks might prove more distressing or dangerous. Second, the burners make a lot of noise and you won't be able to hide them in a distant basement corner. Third, the point of going into the earth is to flatten out energy costs, but by relying on oil and gas you are tied into a system of scarcity and spiraling prices from Day One. You are reducing the economic benefit of the home. Electricity will surely rise in price too, but at least the supply is domestic in some locations and reasonably assured for the near future.

Forced Air. This is the most often recommended option because the moving air helps create ventilation both in winter and in summer. In a forced-air system, blowers push heated air through the house in winter and outside air through the house in summer. The winter heat source can be any type of furnace.

In any case, the heating system used in an earth shelter will be able to function as a backup to the natural heat-buffering abilities of the soil, and will thus be smaller and will work less than any system in a similar-sized surface building.

AIR VERSUS EARTH TEMPERATURE PATTERNS

EARTH TEMPERATURE STATION/1	AIR TEMPERATURE STATION/2	MAXIMUM E	A	MINIMUM E	A	SPREAD E	A	SPREAD DIFF.	JAN. DD (1)	UDD (2)	EARTH SHELTER % HEAT SAVINGS (3)
Auburn, Ala.	Montgomery, Ala.	74	81	56	49	18	32	14	543	279	49
Decatur, Ala.	Huntsville, Ala.	71	81	48	43	23	38	15	694	527	24
Tempe, Ariz.	Phoenix, Ariz.	81	90	59	50	22	40	18	474	186	61
Tucson, Ariz.	Tucson, Ariz.	85	86	65	50	20	36	16	471	0	100
Brawley, Cal.	Yuma, Ariz.	90	95	68	55	22	40	18	363	-93	100
Davis, Cal.	Sacramento, Cal.	76	75	56	44	20	31	11	614	279	55
Ft. Collins, Col.	Denver, Col.	63	72	37	29	26	43	17	1128	868	23
Gainesville, Fla.	Orlando, Fla.	80	82	69	62	11	20	9	220	-124	100
Athens, Ga.	Athens, Ga.	77	81	57	45	20	36	16	642	248	62
Tifton, Ga.	Albany, Ga.	80	83	62	51	18	32	14	400	93	77
Moscow, Idaho	Idaho Falls, Idaho	57	69	37	16	20	53	33	1550	868	44
Argonne, Ill.	Chicago, Ill.	64	75	38	25	26	50	24	1209	837	31
Lemont, Ill.	Chicago, Ill.	65	75	39	25	26	50	24	1209	806	33
Urbana, Ill.	Springfield, Ill.	67	76	39	27	28	49	21	1135	806	29
Urbana, Ill.	Springfield, Ill.	68	76	42	27	26	49	23	1135	713	37
West Lafayette, Ind.	South Bend, Ind.	66	71	38	25	28	46	18	1221	837	32
Burlington, Iowa	Burlington, Iowa	71	77	38	24	33	53	20	1259	837	34
Manhattan, Kan.	Concordia, Kan.	69	80	41	28	28	52	24	1163	744	36
Lexington, Ky.	Lexington, Ky.	68	76	42	33	26	43	17	946	651	32
Lexington, Ky.	Lexington, Ky.	70	76	46	33	24	43	19	946	589	38
Upper Marlboro, Md.	Washington, D.C.	70	77	42	36	28	41	13	900	713	21

EARTH TEMPERATURE STATION/1	AIR TEMPERATURE STATION/2	MAXIMUM		MINIMUM		SPREAD		SPREAD DIFF.	JAN. DD (1)	UDD (2)	EARTH SHELTER % HEAT SAVINGS (3)
		E	A	E	A	E	A				
East Lansing, Mich.	East Lansing, Mich.	63	71	37	24	26	47	21	1262	868	21
St. Paul, Minn.	Minneapolis, Minn.	62	74	34	15	28	59	31	1631	961	41
State University, Miss.	Meridian, Miss.	79	81	55	48	24	33	9	543	310	43
Faucett, Mo.	Springfield, Mo.	65	78	43	33	22	45	23	973	682	30
Kan. City, Mo.	Kan. City, Mo.	66	81	42	30	24	51	27	1032	713	31
Sikeston, Mo.	Springfield, Mo.	71	78	43	33	28	45	17	973	682	30
Bozeman, Mont.	Billings, Mont.	56	73	33	23	23	50	27	1296	992	24
Huntley, Mont.	Billings, Mont.	64	73	36	23	28	50	22	1296	899	31
Lincoln, Nebr.	Lincoln, Nebr.	69	79	39	24	30	55	25	1237	806	35
Norfolk, Nebr.	Norfolk, Nebr.	66	76	40	19	26	57	31	1414	775	45
New Brunswick, N.J.	Newark, N.J.	65	75	42	32	23	43	20	983	713	28
Ithaca, N.Y.	Syracuse, N.Y.	59	73	39	26	20	47	27	1271	806	37
Raleigh, N.C.	Raleigh, N.C.	73	79	52	41	21	38	17	725	403	45
Columbus, Ohio	Columbus, Ohio	65	74	41	30	24	44	20	1088	744	32
Barnsdall, Okla.	Oklahoma City, Okla.	74	82	54	37	20	45	25	1165	341	70
Pawhuska, Okla.	Oklahoma City, Okla.	74	82	50	37	24	45	21	1165	465	60
Corvalis, Ore.	Eugene, Ore.	66	67	46	38	20	29	9	803	589	27
Pendleton, Ore.	Pendleton, Ore.	67	75	39	31	28	44	16	1017	806	21
Calhoun, S.C.	Columbia, S.C.	76	81	52	47	24	34	10	570	465	18

Location	Comparison	E high	A high	E low	A low	Spread E	Spread A	Spread Diff.	DD	UDD	%
Madison, S.D.	Huron, S.D.	61	75	33	14	28	61	33	1628	992	39
Jackson, Tenn.	Oak Ridge, Tenn.	71	78	49	38	22	40	18	778	496	36
Temple, Texas	Waco, Texas	82	86	58	47	24	39	15	536	186	65
Salt Lake City, Utah	Salt Lake City, Utah	63	78	40	29	24	39	15	1172	775	35
Burlington, Vt.	Burlington, Vt.	63	70	5	18	28	52	24	1513	930	39
Pullman, Wash.	Walla Walla, Wash.	60	76	36	32	24	44	20	986	899	9
Pullman, Wash.	Walla Walla, Wash.	58	76	38	32	20	44	24	986	837	15
Seattle, Wash.	Seattle, Wash.	61	65	45	39	16	26	10	738	620	15

This chart shows the enormous heat-saving effect of earth shelter in various parts of the country. It compares high and low temperatures in the earth (E) and air (A). The Spread column shows the range of degrees a house must cope with, and the Spread Difference column notes how many fewer degrees the earth shelter has to handle. The DD columns compare degrees days, a measure of heat requirements; Jan. DD shows how much a surface heating system would have to make up in January. UDD indicates how much the same month would require underground. The last column reveals the percentage of heating requirements earth shelter saves.

ZONING AND CODES

Building codes control the structure and design of various types of construction. Zoning laws regulate the relationships of the buildings to others within the community.

BUILDING CODES

Building codes, more or less, conform to national standards, though there are some seventeen hundred variations scattered among American towns and cities. Zoning laws are established by local governments and vary as much as the complexion of our many city councils.

Anyone who plans to build underground has to cope with codes and zoning requirements that were established with other kinds of construction in mind. In fact, many of these laws make assumptions about the use of subsurface space that bear no relationship whatsoever to today's earth-sheltered designs.

Back in the 1950s thousands of young American couples struck out for the suburbs to start their families in neat suburban homes. Many didn't have enough money to finish their homes in one gulp, so they paid to have the foundations poured and sealed and then moved into these makeshift shelters until they gathered a big enough stake to finish the house. This practice resulted in a rash of laws against living in below-grade basements. Obviously such unfinished, unsafe, unsightly dwellings have nothing in common with the sophisticated earth shelters of today, but the law doesn't reflect that. In some communities outdated regulations like this one have to be met head on and defeated before a family can put its house under the earth.

Similarly, many laws were passed that dealt with underground space as the realm of bomb shelters and bunkers. Rules that were aimed at paranoids trying to escape World War III may stymie a homeowner who couldn't care less about Armageddon and simply wants a satisfying way to save on fuel.

The key to getting around problems with the building bureaucracy is information. You can't figure against the rules that bind you until you know what they are.

Malcolm Wells suggests that when you first begin planning an earth shelter you should go to the local housing permits office and get all the background material you need to put up any kind of house. Don't tell them you're planning a unique structure. Merely get the code book and whatever other printed matter they have handy.

Go home and study the spots off the pages. Again, you may want to get a pro involved here to make sure that your design conforms to all the labyrinthine requirements of the code at first blush. It's easier and more economical to design to code from the beginning than it is to go back and change things later.

When you have a preliminary plan, go to the building code inspector with as engaging a display of your intentions as possible. "When the inspector sees your breathtaking color renderings and charts of fuel savings," Wells reasons, "how can he help but be impressed? Well, there is a way: by having stupid code violations in your design. You won't get anywhere unless you take the code dead serious and work out design problems so the broad intents of the code are followed. Codes are, after all, tremendously helpful in averting most of the disasters which can happen in buildings."

New Hampshire earth shelter builder Don Metz agrees that the best course to follow is to hew to code as much as possible: "I've seen earth-sheltered plans advertised that were in violation of uniform codes," he states. "The people behind those deserve to go out of business. I don't think there is any prejudice [among code enforcers] against earth shelter. I think there is a lot of prejudice against buildings that don't have natural light or that are dark and damp. And well there should be."

Many code rules don't apply to underground buildings because they deal with the wood frames that support surface houses. The difference in materials and techniques used in earth shelters can also sometimes make it possible to gain variances from parts of the code that don't apply to the unique situation.

"Most codes," explains John Barnard, "are designed for highly combustible wood frame houses. I avoid wood and use it only for trim in my atrium homes. My roofs are concrete; my walls are concrete. This is a much more fire-resistant structure than a frame house, so it's possible to get some code rules modified.

"An example is a house I had in Port Chester, New York. Codes require, in most areas, that each room have some method of egress to the outside—a window, door, skylight, something a person can leap through in case of fire. We had bedrooms that egressed onto the sunken atrium. Well, no one was going to argue that the atrium wasn't outside. But the owners wanted to cover the atrium and make it like a sunken greenhouse. Suddenly it was a whole different ballgame.

"But finally we did get a variance so we could cover the atrium, because the house was so much more fireproof than what the code had in mind."

An important word in Barnard's story is "finally." Getting code variances is a long, laborious process that costs money and saps enthusiasm for a building project. Unless you have no choice, it's far, far better to design to code rather than attempt to fight code restrictions. You may win, but you won't enjoy the fight one bit.

In some cases, however, specific code requirements leave an earth-sheltered planner no choice but to fight. Some towns, for example, specify exactly what must go on a roof. If the code says tar paper, shingles and nails, you can't cover your roof with soil unless you get a variance. Some codes are so restrictive of subsurface dwellings that you may either have to claim, as some architects have successfully done, that the house stands "at grade," not below it, or design your home to stand above ground with earth piled around it—a berm design.

"It's very definitely a source of frustration to people who have building plans that don't meet the codes," says Ray Sterling, the director of the Underground Space Center at the University of Minnesota. But, he notes, the situation should improve as the federal government's interest in the new energy-efficient building methods grows and nationwide standards and practices become accepted.

A first step in federal examination of the code problems of underground buildings is the booklet *Earth Sheltered Housing: Code, Zoning and Financing Impediments*. You can get a free copy from the Department of Housing and Urban Development, Washington, D.C.

ZONING

Zoning presents its own problems for earth-sheltered designs. The basic trouble can arise from the simple fact that underground houses are different. Many zoning laws are set up to insure a certain similarity in the structures of a neighborhood; these laws purport to protect the property values of the area by blocking garish buildings or those that serve a different function from the rest of the neighborhood. However, they can also block an earth-sheltered house simply because it doesn't fit into the local scheme of things. It stands out.

On a more prosaic level, zoning requirements for lot size can prove pesky for below-grade development, as can some septic tank rules. But the five big obstacles to zoning for earth shelter are:

Setbacks. Most zoning laws require a fixed setback distance from the property edge to the edge of the house. The setback area brings a certain uniformity to the look of a neighborhood, in addition to insuring access for fire trucks and emergency vehicles, reasonable surety of decent light and air availability to all buildings, and assurance of access to all sides of a surface house for painting and maintenance. Setbacks don't really apply to earth-sheltered houses, though. What if the earth-sheltered house's door meets the setback requirement

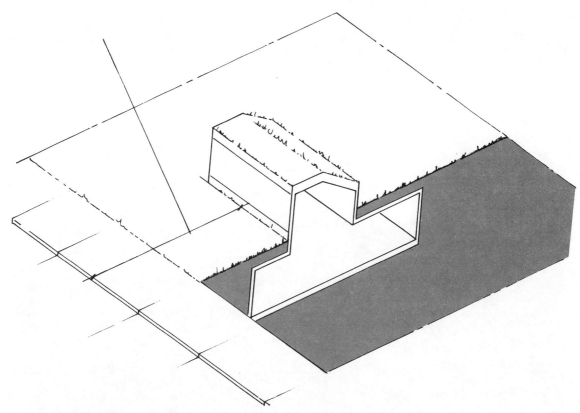

Zoning laws often require houses to be set back a certain distance from boundaries. With an earth shelter, however, it is unclear whether that means the surface exposure or the earth-covered area.

but the subsurface space extends forward into the no-no zone? Does that conform to the law or not? Interpretations vary.

This is an important consideration because it determines how small a lot an earth-sheltered design can fit on. A smaller lot can save thousands of dollars in the eventual cost of the house. In the absence of definitive standards, the setback rules for earth shelters will probably be hashed out over the next few years in court. Best bet: If you can afford to conform entirely to setback requirements, do so; save yourself the aggravation of a long, bitter fight.

Effects of Neighbors. Since building an earth shelter requires digging a large, deep hole, care must be taken that the removal of all that dirt doesn't in any way undermine a neighbor's house. Care in surveying must also be taken to insure that extensions of parts of a submerged house don't wander over a neighbor's property line or come close enough to interfere with the stability of either house. Line-of-sight problems should also be checked. If an earth-sheltered house situated toward the south for solar gain directly faces a neighbor's yard, it might: (1) make the neighbor uncomfortable and (2) expose the earth-sheltered house dwellers to unwanted viewing by the higher ups next door. The relationship of the building to neighboring homes should be carefully considered.

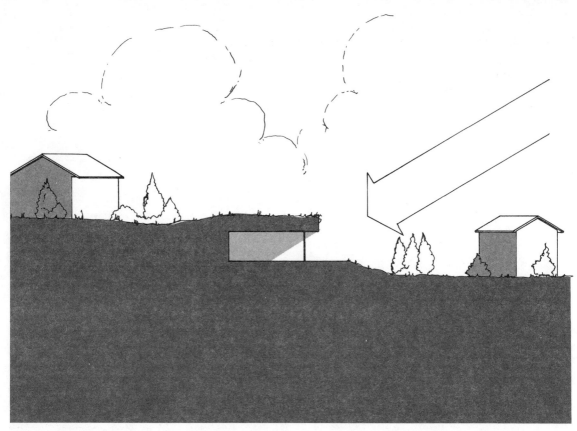

Earth shelter owners may have problems with neighbors. The house in the center faces the sun, and is also exposed to direct view of the neighbor's windows; not much privacy possible here. If the earth-sheltered area goes back too far, it could undermine the foundation of the other neighbor's house or even stray across his boundary line.

Finally, underground water patterns may be disturbed if you're not careful. If water is diverted from neighbors, or if the water table on their property is raised high enough to harm their house, you might be liable.

Solar Rights. You have to make sure that nothing your neighbors are likely to do to their buildings will interfere with your access to the sun, particularly if you are tying earth shelter to solar design. When planning your house, keep in mind that the winter sun sits far lower in the sky than the summer sun. You have to work on the assumption that your neighbor might add a story onto his house and the sun will be at its lowest point in the sky. Will you still get full sunshine? An earth-sheltered house shadowed by a neighbor could become a pretty chilly place. This is a particular concern for earth-sheltered homes because the areas that collect solar heat—either active or passive—aren't nearly as elevated as with surface buildings. This means there may need to be more distance between the property line and the house to assure enough sunlight.

The one circumstance where an earth-sheltered design has the best chance for getting a zoning variance is where a tiny or odd-shaped lot is languishing in the heart of an otherwise thriving area. No surface structure will work on the site, but if rules about setbacks and other details are relaxed, an earth shelter could put the lot to handsome use. The earth shelter can use the subsurface space, with just a small, appropriately designed presence on the surface. City planners might see the earth shelter as a better alternative than a weed-strewn, untaxable wasteland.

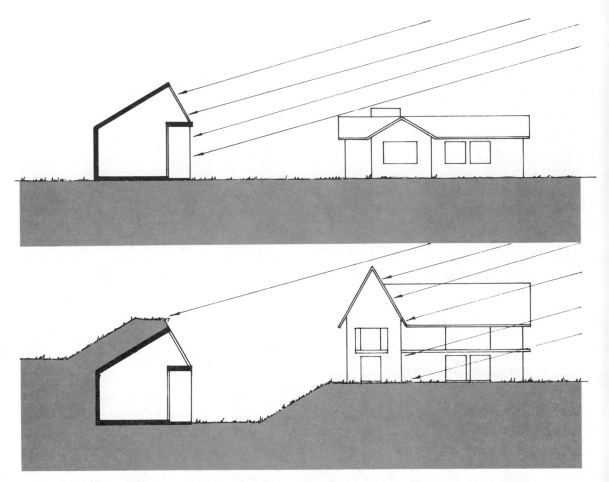

An earth shelter isn't guaranteed access to sunlight that flows over a neighbor's land. Designs should make allowances for sun blockage created by a new building or the addition of another story next door.

COST

How much does an earth-sheltered house cost? The answer depends upon whom you are talking to.

BUILDING COSTS

John Barnard: "Our designs run between 5% and 10% below the cost of a surface dwelling of similar square footage."

Don Metz: "When everything is taken into account, an earth-sheltered design will run about 10% more than a typical home."

Malcolm Wells: "Many underground houses have been built for less than the cost of aboveground ones, but most have cost more. The hefty structure needed to resist all earth loads and the first-class waterproofing on which the success of the house so largely depends tend to offset the savings generated by the lack of exterior finishes on many walls."

Glenn Strand: "There are people who claim to build earth shelters for less than conventional homes, and it probably can be done, but I don't think you have the equivalent structure. That's hard to do. Basically the same type of building and layout as is typical of structures above ground would cost you less. But earth shelters' increased structural costs, waterproofing, drainage, et cetera, will add substantial amounts relative to a conventional structure.

"The range depends on what you're starting with. It could go, initially, from 5% to 40%. You have to increase the structural components to deal with the problems, at the same time eliminating the external finishes, which are difficult things to estimate in the initial plans but add up (painting, for example). Obviously, there are substantial things you can do to make the extra cost worth it."

A conclusion from James Scalise: "When studying the cost of constructing underground structures, it becomes apparent that there are no set figures for cost. This is because of the many variables involved. Some of those variables include:

Excavation. If the soil has a shallow angle of repose (see section on Soil), the excavation will have to be quite large, which raises both the cost of digging the hole and filling it up again. If rock lies near the surface, getting through it will raise the price still more.

Water. If your house dips below the water table (see section on Water), pumps and extra waterproofing will become necessary. If the soil above the water table is particularly juicy, you'll still have to water- and vaporproof like mad. Surface water in large amounts must be countered by land forming and drain building. Water pressures in the ground can add up costs on structural shoring and footings.

Planning and Design. Finding the right site, fitting the house to it, matching structural requirements with your aesthetic needs."

"The best way to save money on construction," says Malcolm Wells, "is to keep things simple. We architects are among the guiltier ones in this regard: We find it hard to simplify, to leave well enough alone. Knowing of so many delightful improvements we can make, we sometimes improve our work to death."

Even if you do keep it simple, the initial cost of an earth shelter, in the view of the majority, will be more than for a surface dwelling, but proponents still feel going down is the best investment because of the long-term returns. They use a method of figuring called life-cycle costs.

OPERATING COSTS

While an earth shelter may cost more to build, it costs far less to live in and operate. Earth shelter offers a "30% operation and maintenance cost decrease" per year, according to John E. Williams of Hanscomb Associates in Atlanta. The savings accrued each year in operation steadily nibble away at the initial savings on a surface building until eventually the underground house becomes the better overall economic deal.

The big savings is, of course, energy costs. A study by the American Underground Space Association compared energy costs over twenty years for a surface home and three similar-sized earth shelters. (Use the specific dollar amounts for comparison only; the study made many economic assumptions that our crazed economy has already knocked into a cocked hat.) The surveyors found that if the surface homeowner paid $3,790 for heat, the cost of earth shelters ranged from $1,516 down to $379. That means underground owners will pay back their excess building costs in ten to sixteen years. The study concludes that "the total cost of ownership is considerably less over a period of years with the earth-sheltered alternative." These savings are lost if you move before the payoff, however. And Americans do seem to be moving with alarming rapidity.

Earth shelter architects point out that the potential energy savings in today's tumultuous circumstances are likely to get bigger rather than smaller, bringing the life-cycle payoff much closer. More pessimistic souls note that even if the building costs a bit more up front, at least you know you'll be in better shape when the oil runs dry or the embargo comes down. You'll be able to meet your heating needs with passive solar or a simple wood-burning stove—something you can control yourself and not have to fear.

They also argue that costs for earth-sheltered construction will tend to drop —or at least hold their position in the face of a rising tide of construction costs —as designers get better and better and mass-produced materials specifically intended for below-grade dwellings become more common. Today everything for earth shelter is either adapted or custom-made. Problems must be individually solved. There is as yet no such thing as a book of complete engineering tables for earth-sheltered situations. The novelty of earth shelter adds to its cost, and as that novelty decreases so will the cost.

According to the American Underground Space Association, "It appears that the cost of earth-sheltered construction is quite comparable to good-quality conventional above-grade housing . . . Once contractors have gained some experience with this type of construction and can confidently predict their own costs, the price is quite likely to go down."

FINANCING

In today's high-interest, unstable, tight-money market, any kind of mortgage is a chore to find, but nailing one down for a home with an innovative design can be a long, exasperating chore.

BANK LOANS

Bankers are negative people by trade, breeding and inclination. Given a reasonable chance to say no, they will. They shy away from anything that hasn't, in their eyes, withstood the test of time. Even worse, they are inherently suspicious of the personality of a customer who would be interested in trying something that hadn't been around for centuries.

The greatest struggle you may face in establishing an earth-sheltered life is finding the money to pay for it all. You have to fight against the newness of what you want to do: the strangeness of design, which may not appeal to banker sensibilities; the engineering problems and benefits, which may be too complex for the banker to fully grasp (especially if he or she doesn't feel like working at it); extreme competition from people interested in building "normal" houses; worry by the banker that your house doesn't belong in this community; difficulties experienced by the bank in determining what your house is worth.

This is a heavy burden to clear away before you can get your house under way, but it is not impossible to overcome. Mortgages for earth shelters may be harder to find than those for surface homes, but they aren't impossible. Thousands of subsurface buildings have been built, which means that somehow thousands of people managed to find financing.

Happily, the financing situation for earth shelters isn't as bad as it was for the real pioneers of ten years ago. They were often considered crazy troglodytes. By now widespread press coverage, conferences, *Architectural Digest* articles, TV reports and other forms of dissemination have generated a wider awareness of what earth shelter is and why people are becoming more interested in it. Most bankers today will at least have heard about earth shelter, even if they don't know much about it.

Still, the typical banker is likely to dismiss the movement as a fad that has captured the attention of reporters but won't last. When it dies out, the banker won't want the bank's money dying with it. You will probably have to overcome this prejudice—even today.

You have to keep in mind, too, what bankers have foremost in their thoughts. You have to look at your situation from their point of view, so you can try to answer their objections before they make them. To bankers one thought is paramount: They cannot let their bank lose money on a loan. Their career depends on it. They won't loan money when they think the chance of losing part of it is too great. In fact, they will need to be convinced that losing any part of the loan is virtually impossible. Only then will they warm—slightly—to the idea.

Bankers look at two overriding factors: you and the house. If they are convinced that the house is a no-loss proposition—as in the case of a fully leased apartment building—they can loan the money with confidence that the bank can't lose even if the borrower turns out to be a deadbeat. As their skepticism about the house grows greater, their scrutiny of the borrower's finances and personality will grow more stringent.

You will have to give the impression that you are the kind of person the bank should be dying to lend money to: solid, responsible, energy-conscious, patriotic, dependable, on top of things. You'll also have to prove to the banker that you have more than enough money or assets to cover the investment you're making in the house.

All this may not seem fair, but it's a reality. You'll have to confront this kind of situation if you're going to get money for your house. In short, you have to sell your house to the banker. You're not begging. You're making a presentation that is so overwhelming in its facts that the bank should have no option to turn you down. If your presentation is strong enough, eventually you'll find somebody who'll agree with you.

Where to start? Begin your money hunt with banks and savings and loan associations with whom you do business. The more the bankers know you—assuming, of course, you've been a good customer—the more relaxed they are likely to be. Even if these folks turn you down, they provide a relatively friendly audience on whom to practice your presentation.

If you have an architect or engineer involved in the project—and you ought to—use that individual's connections. Some builders and designers have hand-in-glove relationships with finance institutions. They may be willing to take a risk on the architect that they'd never take with you all by yourself. In one notable case, John Barnard, the New England earth shelter pioneer, is on the board of a local savings and loan association that says it will back him solidly. Such financing clout is one pleasant offshoot of getting pros involved in creating your house.

When you approach a bank, you should be prepared. "Mortgage officials are not known for their enthusiasm for any kind of unusual building," notes Malcolm Wells. "You'll have to find the few who are. Impressive presentations of your beautiful house drawings and energy-saving calculations help. Always emphasize the value of permanence. Your underground building—which will be, of necessity, built to last—will still be young when mortgage-burning time comes."

You should have done your homework. You should know the answers to all likely questions. You should have specifics on your costs, savings, materials, contractors and every other aspect of the project you can think of. You should have all pertinent local codes and permits firmly fixed in your mind and know precisely how your house conforms to each.

Drag in evidence of how widespread and successful earth shelter has been elsewhere. Buy books on the subject. Order piles of pictures and plans from earth shelter architects all over the country. Solicit testimonials from earth shelter dwellers or, better yet, from their bankers. Have everything on paper; bankers like paper, they can hold it, rustle it and show it to the boss. Tangibles mean a lot to bankers.

If your presentation gets a cold shoulder, politely ask what deficiencies in your plans or presentation kept you from getting the loan. Thank the banker for his time. Use this individual's suggestions to improve your next sales pitch.

As a rule, by the way, you should try to get to see the highest-ranking bank officer you can. The underling is just learning the rules and is not likely to bend or break them. The boss, however, is probably smarter, more confident and more flexible, and might be itching for something a little new and different to spice up the old portfolio.

Another thought to bear in mind when confronting bankers: They are looking at a mortgage to you as something they can sell. The bank won't hold your loan note while you pay off the house. They'll sell it on something called the secondary mortgage market. The bank will get less for the loan than the total of your payments, but they get the money today, not in dribs and drabs. Finance institutions want mortgages they won't have trouble unloading. You should make sure to address this point, since it shows you know their business and gives you a chance to deflect their secret objection before they can make it.

If your personal bank doesn't go for your pitch, ask around among bankers, real estate people and builders you know to see if any bank or banker in town has a reputation for being a little looser than the norm. Best of all, find out if any institutions in the area have already backed an earth-sheltered building. Try this list second.

FEDERAL HOUSING ADMINISTRATION LOANS

If you still haven't gotten anywhere, maybe you should try the Federal Housing Authority (FHA), the government-backed home lender. In a perfect world FHA would be leading the way in loaning money on energy-saving, long-lasting ecological homes. The government could provide the wedge that would break open the mortgage market to these sensible designs. But, predictably, FHA and the Farmers Home Loan Administration, another government-backed lender, have lagged in supporting earth shelter.

A national survey by *Earth Shelter Digest & Energy Report* in the spring of 1980 found that responses of local FHA offices to earth-sheltered ideas followed no set pattern. Views differed from office to office—even among different workers in the same office. An individual at the Boston FHA office, for instance, declared: "We have no plans to get involved (in earth shelter). I know of no way whereby an earth house would have a market in New England." Incredible! This from a man whose office is in the heart of the most active earth shelter region of the country. Malcolm Wells and John Barnard practice on Cape Cod. Thomas Bligh teaches at MIT. Don Metz and Earthworks do their underground thing in New Hampshire. And there's no market for their houses?

At the other extreme is the Helena, Montana, office, which has approved a couple of earth shelter designs. Art Clingan, the chief architect there, is interested in structural integrity, not tradition. "We don't want someone sitting at his kitchen table sketching and thinking the plan could carry the load," but if the house makes engineering sense his office is open to it.

So you can anticipate a certain level of ignorance and resistance from FHA if you aren't lucky, but you shouldn't let it get you down. If one officer dumps your plans, try another. If the nearest office doesn't help, try the next nearest. Ask your representative in Congress to kick the FHA sluggards in the bureaucratic butt. Their resistance is arbitrary and unnecessary. No rules block FHA from financing earth shelter. Make sure they know you know that. They have approved concrete slab on grade buildings, so why not a little below grade? Make sure you let them have that, too. And point out that their fellow officers elsewhere can't understand resistance to earth shelter. "I guess there are offices of ours that don't want to work with earth shelters," says one FHA field officer, "but I can't understand that."

The ultimate trick to getting earth shelter financing is to be tenacious and smart as a fox. Thomas Bligh, for instance, has developed "a couple of twists" that he has found work quite well in disarming bankers. "You don't go rushing into a bank saying, 'Look, I want to build an underground house.' That's scary enough that he'll throw you out the door. You go in and say, 'I want to build this exciting house.' You show him the plans and talk all about the energy savings. He says, 'What's all this soil doing up on top here?' You say, 'That's added insulation, keeps the weather away.' You sell it as energy savings and nothing nutty at all."

As energy costs keep rising and earth shelter becomes even better known, the resistance to lending will slowly fade. "There is a lot of inertia still," notes architect Glenn Strand, "but you have to look at the situation from the bank's point of view. It's a bit hard to keep your perspective on it. If somebody wants to build something, and he starts pounding on banks' doors, to him it's really painful that they're not willing to explore and take risks. But at the same time, if you take a long-distance perspective, you can say 'Maybe that's an appropriate response. Maybe we've gone so fast in developing this that we have to give them a little time to catch up with us.'"

And, in the cautious-banker fashion, they do seem to be getting the idea: "With earth-sheltered homes having the potential to reduce energy costs 30%, 50%, even 80%," states Mark L. Korell, a former executive assistant at the Federal Home Loan Board, "they should enjoy special favor as a way to reduce the risk to lender and owner alike" of overloading the family budget with escalating energy expenses. In that view, earth shelter presents a far lower risk of default than regular homes; energy costs are far less likely to strip the family budget of the money earmarked for the mortgage. Make that point with your banker.

As you trudge from bank to bank to bank, remember that you are not alone. Others have been here before you and they've been successful. "The lesson of the many underground buildings completed or under construction across the country today," writes Malcolm Wells, "is this: legal matters and financing do not stand in the way of underground architecture—if you are willing to stick with the project and guide it past the obstacles. In many areas the building of anything is delayed by a long and tedious process . . . Often these delays occur not because you're building underground but simply because you're building."

GETTING FHA'S ATTENTION

This is how one family wangled an FHA-insured loan on their earth-sheltered house.

Winter 1977–78	We drew up plans for the house and talked with our bank, the Bozeman Federal Credit Union, about a construction loan and a long-term mortgage.
April 24, 1978	We received approval of a construction loan.
May 8, 1978	The papers for the construction loan were signed and we received a verbal commitment on a thirty-year mortgage.
June 6, 1978	We started construction on the house.
January 13, 1979	We moved into the house, although it was only 75% completed.
January 16, 1979	We asked the bank to convert our loan but they refused, saying the house had to be 100% completed.
February 6, 1979	I was told by the bank that interest on long term financing had gone from 9.5% to 12%.
March 1979	We met twice with the bank manager to find out what our situation was and how to best solve it.
April 1979	We stated our case before the board of directors of the bank, but they informed us that they were not going to honor their verbal commitment to finance us for thirty years.
May 8, 1979	The bank gave us a four-month extension on our construction loan.
June 1979	We applied for an FHA-insured loan through the First Federal Savings & Loan Association in Great Falls, Montana.
	The house was appraised by the local FHA appraiser. She was accompanied by a representative from the state FHA office in Helena.
Late June 1979	Don Inman, a cost analyst from the state office in Helena came to inspect the house.
Middle of July 1979	Mr. Kafemtzis, with the architecture department of the state office, came to inspect the house.

First week of August 1979	We received a denial from FHA.
Second week of August 1979	Because FHA seemed prejudiced, I did not think that it would do any good to appeal their decision, so we called Rep. Pat Williams and Sen. Max Baucus and asked for their assistance.
August 13, 1979	Pat Williams held a meeting in Bozeman and we presented our case to him in person.
August 14–21, 1979	We kept in touch with Pat Williams and Max Baucus by telephone.
August 21, 1979	We decided to make the trip to New York City and Washington, D.C.
August 22, 1979	I called the state FHA office and spoke with Mr. Kafemtzis. I asked him to reopen my application but he refused.
August 24, 1979	Because of our trip east, we began talking with Judy Chapman in Pat Williams's office in Washington, D.C.
August 27, 1979	We met with Brian McGroarty and he took us to the office of *Earth Shelter Digest & Energy Report*. We talked with the publisher and the editor and they advised us to talk with the people in the FHA office in Minneapolis.
	We met with Walter L. Kroenke of the appraisal department in the Minneapolis area FHA office.
August 31, 1979	We talked with Judy Chapman by phone and she informed us that she had made an appointment with HUD for us in September 1979.
September 4, 1979	We met Judy Chapman and she went with us to the HUD building.
	We met with Carl Crowe, a program specialist, but he was unable to give me the assurance I was seeking. He then took us to his superior, Mr. William Halpern, director of Single Family Development. Mr. Halpern called the regional office in Denver and discussed the situation with them. We were then told to go back to Bozeman and reopen our application with FHA.
September 5, 1979	We met with Rep. Bruce Vento of Minnesota.
September 19, 1979	We sent in the request and supporting evidence

for FHA to reopen our case to First Federal Savings & Loan Association. In the meantime, Pat Williams and Max Baucus both sent letters directly to FHA in Helena, encouraging them to reconsider our application.

September 25, 1979

Mr. Marxer, director of FHA in Montana, and Mr. Clingman inspected the house.

September 26, 1979

I talked with Mr. Clingman on the phone and he seemed concerned about the drainage in my driveway.

October 2, 1979

I talked with Mr. Marxer on the phone and he mentioned that my loan was being coordinated with the regional office in Denver.

October 11, 1979

Mr. Marxer informed us that the house had been appraised at fifty thousand dollars and that the loan had been approved.

October 19, 1979

Written approval was received from FHA with three special conditions.

November 21, 1979

I called FHA and informed them that the house was completely done.

November 29, 1979

Jerry Boone of the FHA office in Helena inspected and approved the house.

December 6, 1979

We received a call from the FHA office in Helena assuring us that everything had been approved and that the paperwork was being sent to the bank. (We switched to Empire Federal Savings & Loan Association of Bozeman because they had some low-interest money available.)

December 18, 1979

We closed our loan with Gary Temple, a loan officer with Empire Federal Savings & Loan Association.

INSURANCE

Why are some thirty earth shelter homes on the drawing boards or in the dirt of Oklahoma right now? Because they can't be blown over by that state's frequent, violent tornadoes.

Why does Jay Swayze sell cavern homes to wealthy sybarites of Las Vegas? Because they are the most secure form of housing possible, as easy to break into as a bank vault.

In other words, there are good, strong reasons why earth-sheltered design should make an insurance agent smile. Living in the cradle of soil offers a range of insurance benefits no other form of housing can match. Underground homes

- survive earthquakes, tornadoes and other natural nastinesses better than surface dwellings.

- are made of materials that don't burn. These homes are buried in dirt, which squelches flames. Their controlled ventilation would slow or smother the spread of a fire.

- don't suffer frozen, burst pipes because all pipes are insulated by the ground. Heavy rains won't hurt them because they are thoroughly moisture-sealed from the first moment.

- Thieves, if they can find a home shrouded by soil and shrubbery, face the daunting prospect of having limited means of escape and access.

- Even bombs might leave an earth-sheltered house—with its thick walls and pressure-conscious design—intact long after surface buildings have collapsed.

Given all these points, you'd think insurance rates for earth shelters would be way below those of surface buildings. Unfortunately, this is not yet the case.

Insurance rates are based on actuarial studies, statistical compilations of facts about performance over long periods of time. These tables allow insurers to set rates that protect them from excessive loss.

Since below-grade living hasn't been around very long, insurance companies haven't yet compiled enough data to formulate separate rates for the new form of design. Today, at least, an earth-sheltered house is just a house to most insurance agents.

In the absence of hard facts to the contrary, insurance agents will make assumptions that are just about as pessimistic as those of bankers. They will assume the worst possible performance by earth shelters until actual circumstance proves them wrong.

This means insurance rates for underground construction haven't been significantly below those for other buildings. However, that is slowly beginning to change. Two insurance companies, Safeco and State Farm, are toying with the notion of offering significant rate reductions to below-grade dwellers, and others are certainly entertaining similar thoughts.

An earth-sheltered design won't save you insurance money immediately, but in a few years, while your neighbors' rates are going up, yours will almost certainly be coming way, way down.

The one area of insurance where living below grade does become a drawback is liability. People can walk off the end of your surface-level roof. They can trip and tumble down your excavation. They can run into your hill-like home with their dirt bikes. Because an earth shelter is so much less obtrusive than surface buildings, it is more likely that it will not be noticed by wanderers and might endanger careless visitors. To get the best insurance deal, you should make sure that these drawbacks are countered. A low fence around the roof might be a simple solution. Or perhaps a small spotlight pointing toward the house at night so that accidents will become less likely.

PSYCHOLOGY

What does it feel like to live beneath the earth? What is the psychological impact of being surrounded by, rather than on top of, the ground? The answer depends upon which source you turn to and what kind of underground environment is studied.

Experts agree that the most alienating underground environment is the enclosed, windowless cavern. In 1974 psychologist Robert Sommer examined the reactions of employees working in an underground, windowless office. They complained of "stuffiness and stale air, the lack of change and stimulation, and the unnaturalness of being underground all day." Other underground workers felt that they lost their sense of time; their brains felt heavy and slow.

However, workers in Kansas City's huge underground industrial complex generally seem to like their surroundings; they comment favorably on the quiet, which they find relaxing, and the year-round comfort of the earth-buffered atmosphere.

Even in a totally windowless underground environment, however, studies have shown that no behavioral problems or learning difficulties cropped up. At the Abo School in Artesia, New Mexico, which was originally designed as a bomb shelter, a psychological survey found that "after ten years of experience with children attending an underground and windowless elementary school, the professionals concerned with the health care of the children—are generally convinced that not only is the school not detrimental to the physical and mental health of their patients but it is actually a benefit to some. . . . Nine out of ten [members of the community] recommended that other schools be built like Abo if such schools cost no more to build than other schools." And a University of Michigan Architectural Research Lab study found similar results at other underground classrooms.

Even in the absence of windows, earth-sheltered environments can be open and pleasant. The key is creating a variety of light effects.

The ability to work at peak efficiency and happiness in a total underground environment seems to be linked to the sense of space and openness that can be designed into the place. Jay Swayze has had some success with his massive caverns when the expanse has been large enough to give a solid sense of there being an inside and an outside, even if both are enclosed.

Perhaps the most spectacular instance of such a design is the Kocho Restaurant, which takes up four basement floors of the New Yuraku office building in Tokyo. One steps off the elevator into a vast Japanese garden with several buildings scattered about. The visitor, according to an article in *Japan Architect*, "enters a different world where shackles of time and place are forgotten. Everything here is artificial. The entire restaurant is so cut off from surrounding walls, and even from the gray concrete ceiling two stories above the garden floor, that the illusion of having stepped out of a large city into a rural retreat of the past is completely successful. Lighting, air conditioning, even the recorded sounds of the insects and of flowing water, dramatically heighten the effect."

One tool that can bring ambience to utterly underground spaces is psychic lighting. Jay Swayze includes a lighting system in his homes that changes the intensity and quality of light analogous to the sun passing overhead. Faber Birren, an expert on light and color, has stated that such psychic lighting could make the most remote underground areas habitable, even enjoyable. "In an office, factory or school, daylight sources plus some ultraviolet may be utilized for a good part of the day. For psychological and emotional reasons other light in other intensities and tints may be programmed: warm light in the morning, increased intensity and whiteness as the day progresses, 'complexion' lighting at coffee breaks or during the lunch hour, pink or orange at dusk."

Of course, most earth shelters won't be totally wrapped by soil. They will have significant contact with the surface; many will even lie above the surface. But the insulating effects of earth create an internal environment different from that in a surface dwelling.

According to James Scalise, "A building submerged into the earth experiences less interaction with external stimuli [than a surface dwelling]. As the climatological effects flow from the atmosphere to the ground, the earth absorbs their effects, acting as insulation. The building fabric becomes relatively unaffected in comparison with a fully exposed above-ground building. The stimulation from sun, wind, and precipitation becomes subdued as the actual building skin becomes the thickness of the earth."

In other words, an earth-sheltered home is quiet in a way no other home can be. Underground homeowners, as a result, tell tales of huge shifts in their awareness after they pulled the blanket of earth over their day-to-day lives. "When we first moved in here," says Barbara Webb, an earth shelter dweller in Missouri, "it was so quiet at night that I had some trouble sleeping. But as soon as I got used to it I slept better than I ever had before in my life. Now, when I go to visit other people, I have trouble sleeping at their houses because it seems so doggone noisy all the time."

So quiet are earth-covered homes that Andy Davis, an earth shelter builder and franchiser from Illinois, tells a tale of the day cows broke a fence and charged over his roof. Down under he never heard a thing.

The silence means that earth-sheltered homes near factories or highways or airports are far more pleasant than their surface-level counterparts. Malcolm Wells argues that all the homes in the paths of big jets should be earth sheltered. "Instead of holding their ears everytime a 747 goes over, these folks could be happily going about their business underground."

In addition to silence, the quality of light in earth shelters is different from that on the ground. It tends to be more directed—streaming through the glass-filled south wall but merely trickling through skylights and small northern windows. Instead of uniform brightness throughout the house, brightness levels can vary; the sense of the sun passing overhead is much stronger than at the surface.

Comfort in an earth shelter seems linked not so much to the number of windows but to the sense of visual contact they bring to the outside environment. Homespun builder Mike Oehler of Idaho handled this problem by designing a house on various levels, with narrow clerestory windows designed to lie at eye level on each landing. No matter where in the house you are, you can always see outside. This makes for contented earth dwelling.

Skylights can also increase contact with outside light, but they must be carefully designed so that they don't lose too much energy in the cause of brightening the house. Flat or bubble-type skylights lose energy by the bushel. But directional skylights angled toward the south can reflect light down into the house, garnering heat from the southern sun. A carefully calibrated overhang can allow winter sun to reach the skylight while shading it from the higher summer sun.

Other factors besides window space and placement can affect the impact of light inside an earth shelter. Scalise notes that "painted walls and ceiling should furnish reflecting surfaces for light and should be pleasing to their created environment. High levels of illumination and white walls can physically destroy the eyes. Tints are preferred to saturated colors, and are also preferred in large rather than small areas. Light colors are active, while deep colors are passive. In a given space, deep orange will have the most exciting influence, yellow-green the most tranquilizing, while violet is the most subduing. As a rule, primary colors are strong in impulse and have universal appeal; intermediate colors are more subtle and liveable. Primary colors elicit a direct, impulsive reaction, and

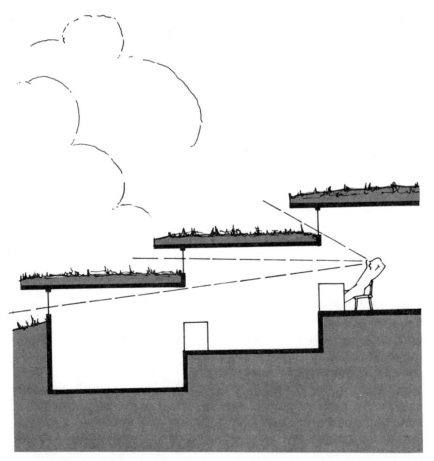

Clerestory windows designed in relation to different house levels can provide constant views of the outside without exposing much of the house's surface.

when used in interiors occupied for long periods of time they are likely to be somewhat monotonous. If the average person is to be emotionally pleased, it may be best to use simple colors: greens, blues, pinks, yellows, grays. Few people like hues of purple, orange and yellow-green as a 'choice of heart.' These natural predilections can be important to the viability of a space, especially one used for trade."

Not just the amount but the quality of light can also affect the emotional impact of a submerged space:

Different qualities of light, even daylight, create different moods. Light of blue, cloudless skies with unobscured sunlight is animating, cheery and exciting. Light of overcast skies is fatiguing, oppressive, and subduing. When a partial eclipse of the sun occurs under clear skies, daylight seems pale and depressing. The theater makes use of colored lights to create moods. According to Dr. Robert R. Ross, a drama professor at Stanford University, gray, blue and purple hues create tragedy; red, orange and yellow hues create comic atmospheres. William A. Wellman, another drama teacher from California, refines this even further: "Red for vigor, yellow for warmth and joy, green for abundance and health, blue for spirituality and thought, brown for melancholy, gray for old age, white for zest and awareness, and black for gloom."

James Scalise lists the lighting guidelines that he feels no earth shelter design should overlook. A good design should

- plan for sufficient light to convey all the illumination human beings need to work effectively at their tasks and to be fully alert to their surroundings.

- work with familiar, rather than strange, elements in illumination and color to avoid confusion or emotional rejection.

- never forget the factor of human appearance. People are being dealt with, not just a pair of eyes or isolated tasks.

- Give all elements in the interior a realistic appearance without flattening them out with lighting or distorting them unfavorably with color.

- Plan for variety as opposed to monotony. All details of an environment should be revealed distinctly and not lost in uniformity.

- Hold light levels and color contrasts within the limits of reasonable adaptation.

- Take any liberty that is appealing, but do not force the human organism beyond its normal and agreeable capacities.

Interior decoration can also have enormous effects on the quality of light and ambience inside an earth shelter. Wall colors can make rooms look bigger and brighter. Transparent Lucite room dividers can organize space without limiting the spread of light throughout the house. Brightly colored plants can give a place a sense of airiness and outsideness that nothing else can. Fish tanks, terrariums, indoor rock gardens, bright wall hangings (especially slightly reflective ones), mobiles and furniture can all contribute to the comfort quotient of an earth-sheltered interior.

Finally, artificial lighting can create visual interest around the house by varying the degree and quality of light produced. Different intensities and colors of electric light and the mixing of neon and fluorescent light can combine to keep the interior exciting to the eye. A bland sameness of light everywhere in the house bores the eye and numbs the brain. Light level can also generate a subtle sense of place, denoting different areas within the house in the absence of walls.

Hard evidence of long-term reactions to living in an earth-sheltered environment aren't available yet because the movement is so new. But the preponderant opinion today is that a well-designed subsurface space can be every bit as ambient, pleasant, exciting and psychologically balanced as its surface counterpart. It is, in the words of Missouri underground engineer Gordon Moore, "a natural way to live."

One properly placed window can suffuse several rooms with light. Earth shelter doesn't require a shadow box existence.

Interior decoration can create a bright environment whether the ground surface is in view or not.

PART THREE

INSIDE EARTH SHELTER HOMES

JOHN BARNARD'S ECOLOGY HOUSE

One of the earliest of the current crop of earth-sheltered designs, John Barnard's Ecology House rests on a wooded lot near Marston Mills, Cape Cod. It is in the classic atrium design, being perhaps the first building to show how earth sheltering and the requirements of modern American life could be melded into a single structure.

Barnard got the idea for his house while on a trip to Pompeii. He and his wife, Barbara, were sitting at a Romanesque-style cafe situated around the traditional atrium. Suddenly Barnard realized that the internal nature of the atrium design allowed light to enter the area, even if it was blocked from the street. The same thing, he thought, should hold true if the entire square were sunk into the earth.

Barnard's father, likewise an architect, had spoken with him as a boy about the possible benefits of surrounding buildings with earth. "When I was six or seven," Barnard recalls, "my father said, 'You know, son, if a house were built deep enough in the earth, it could be 60° all year round. Why not build houses down there?' The reason was light, was making the houses be not only efficient but something people would want to live in. The atrium concept accomplished that."

His wife's response to Barnard's brainstorm was succinct: "I hope you and the mole you marry have fun down there, but I'm not going." Eventually she changed her mind.

"I like to be outdoors more than in. So at first underground didn't interest me. But as far as I'm concerned, with an atrium like this, it's ideal. I love it," she says. As for heating, Ecology House uses rooftop solar collectors as well as standard electrical resistance heating. Even without the heat pump Barnard plans to install, he saves 25% on his fuel bills. He also predicts these savings could be increased to 75% with a transparent shield over the atrium.

You enter the result of Barnard's brainstorm by descending a stairway that curves into the center of the gardenlike, three-hundred-square-foot brick atrium. Around you are the glass walls of the house's rooms. Each room except the bathroom and the laundry have access into the atrium. Each room receives generous amounts of light from the central opening. The house, as a result, has the airy, bright quality of a hilltop cottage rather than the dingy aura of a cave.

The quality of light is a bit different than on the surface, however. It changes texture steadily as the sun passes over the south-facing atrium. The shadows move more clearly in relation to the direction of the light, and the beginning and end of each day have a long-lit, twilight quality because the sun lies below the horizon of the lowered house.

Floors in Ecology House consist of three and a half inches of reinforced concrete. The sheltered walls are also composed of poured concrete, with Styrofoam panels on the soil side for insulation. Barnard's roof is made of precast concrete planks, a form some architects have doubts about but which he feels he can use because the house lies relatively close to the surface.

Barnard waterproofed the roof with three plies of sixty-pound asbestos felt topped with hot pitch. Pitch alone was used on the walls.

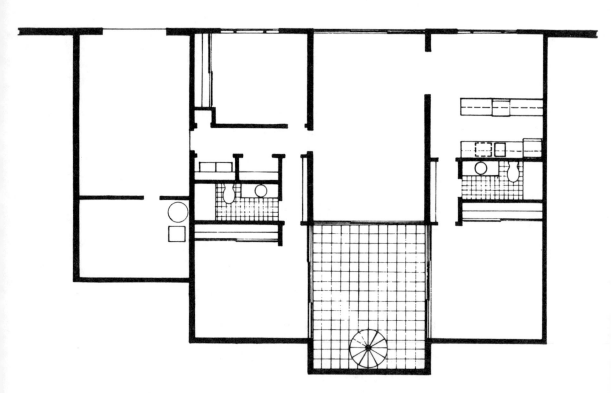

Floor plan of John Barnard's Ecology House. The checkered area is the sunken atrium. The pinwheel in its center is the stairway.

THE CLARK-NELSON CULVERT HOME

The house looks like a pair of tired giant eyes peering out at the forest. Two semicircular sections rise from the hilly landscape. If it weren't for the glass, you could easily imagine them as unusually symmetrical natural hills. But they are the exposed supports of a 2,500-square-foot residence that uses a design concept unique in American architecture. They are storm culverts, cut in half lengthwise and buried.

"The design of the structure came to me quite by accident," says architect Michael McGuire, "while we were out on a picnic at a state park. After the picnic I wandered down a path in the woods and came to a stream. As I looked up the hillside, there was this arched steel culvert spanning the stream, with a roadbed over it."

The culverts were designed to withstand enormous loads, so McGuire reasoned they were naturals for earth sheltering. "They offered all the advantages of a traditional sod house," he states. And the culverts were far cheaper and easier to build with than concrete and steel.

McGuire made the house for Pat Clark from culverts thirty-eight and fifty-eight feet long, linked by a fourteen-foot hallway. The living room, dining room and kitchen lie in the shorter tube. The longer tube holds a bedroom, a studio and a separate apartment for a caretaker's family. Windows and doors at the ends of the tubes let light in and let inhabitants out onto small patios. Laundry and mechanical systems dwell in the linking hallway.

Building the home required little more than pouring footings and then attaching the culverts to them. The culverts were heavily waterproofed with fibrous asphalt and plastic, then backfilled with sand and buried.

Insulation was in the form of polyurethane foam that was sprayed on the inside walls and painted, so the walls themselves are the insulation material.

Heat for the home comes from fireplaces—there is one in each room—and an oil backup furnace. Owner Pat Clark claims that "we pay one third of what we would in a regular structure" for heat.

The following are some other features she likes about the house: "Security is one of the main things, and you don't hear a lot of noise down here. I also feel a sense of real comfort in the lines of this structure. I feel I'm part of the environment."

The design does have its drawbacks, though. It's extremely dry, even with the bath and shower running inside. Apparently, the polyurethane walls are thirsty, and retain moisture for a long time.

However, says architect McGuire, the house "is comfortable in winter and summer (without air conditioning) and there is no sign of the dampness or basement effect one could expect during humid Wisconsin summers."

The Clark-Nelson culvert house shows just how simple an earth shelter design can be and how comfortable even an unorthodox structure can be in the hands of a creative architect. Solving the technical problems of living with the earth doesn't have to be a terribly complex process. And even the most rudimentary solution can be perfectly liveable.

An earth shelter made from sewer culverts. This design for a Wisconsin home shows how simply earth shelter problems can be solved by creative architects.

BARBARA AND LARRY WEBB'S SIMPLE ELEVATIONAL HOME

Barbara and Larry Webb were a typical young couple in Drexel, Missouri, living in a typical young couple's "little modern house." Then, one day, they saw an article in a magazine about a new type of house, one that was sheltered by soil so that it used far, far less energy to keep warm.

"As soon as I saw it I thought it was something we should think about," says Barbara Webb. "It seemed to us, with the energy crunch on and probably going to get worse, we should do our best to hold the line."

Filled with enthusiasm, the Webbs gathered together all the information they could find on underground construction. Then they drew up their own set of plans for a simple earth shelter. For weeks the couple carried their homespun design from bank to bank, with no success whatsoever at getting financing. One of the big drawbacks to the bankers was that the plans had not been approved by a professional engineer.

So, chastened and eager to get started, the Webbs gave in and went to a structural engineer. He gave the plans a final once over and convinced Larry Webb that a pro's contribution was necessary. "It is important to have an engineer design the structural part of the home, since it will be holding up a lot of weight."

Armed with the new plans, the Webbs returned to the money market and again got banged around pretty soundly. Finally, though, a small local savings and loan association came through for them. The Webbs had all this trouble, it should be noted, with a 50% down payment in the bank.

Once they had the money, the Webbs got a building contractor to actually put the house up. The entire structure consists of poured concrete with six-inch-thick walls and a ten-inch ceiling. It is heavily waterproofed and insulated. The heart of the 11,250-square-foot home's design is a pair of atriums that provides additional access to the outside and brings light to all parts of the house. Every room except the laundry has a glass wall that opens onto a courtyard. The living room even has two doors at opposite ends.

At first sight, the Webb house looks like a grassy knoll with a barbecue pit on top. The pit is actually the top of a fireplace chimney. From the top of the knoll you can see the atriums and the steps descending down toward their centers.

The house has been a resounding energy-saving success. "We used one hundred kilowatt hours less electricity this winter than we did in our earlier home," says Barbara. "We've used no air conditioning at all in the summer and remained perfectly comfy."

And it hasn't proven uncomfortable to live in—an early concern of Barbara's. "I used to like to look out the windows and see the neighbors and the street. Here I can't do that. But I don't miss it so much because I can't hear it either. It's much quieter down here. My friends and relatives had a really funny feeling about our doing this at first. They couldn't understand why we wanted to live in a root cellar. But now that they've been here, they like it. They feel comfortable, not like they're underground.

"That's probably been the big thing for me. The house doesn't feel like it's under anything. It just feels like a nice, comfortable house," However, Barbara admits that the place does have one drawback: "Now I have to carry the groceries downstairs instead of straight in the door."

Despite Barbara's lighthearted tone, it should be clear that they went about the construction of their earth shelter in a most serious fashion. Despite Larry's desire to be integrally involved in the design of his new home, the couple brought in a professional contractor for the crucial job of pouring the structure. And they checked all their plans with engineers. If they hadn't been so careful in the planning and construction stages, they almost certainly wouldn't be as carefree in the living stage as they are.

The exterior of the Webb home. Light enters through the glass roof and through strategically placed windows on the south-facing walls.

Larry and Barbara Webb relax in the front room of their simple earth-sheltered home.

DAVID WRIGHT'S SUNDOWN HOUSE

Environmental architect David Wright has snuggled his passive solar home, Sundown, on a rolling hill above the Pacific Ocean, along California's gorgeous rocky coast. A classic inset design, Sundown tucks smoothly into the face of a slope, virtually invisible from behind, open in front to the sun and the view.

The house stands in two sections that front an informal, barely subsurface-level courtyard. In the larger twelve-hundred-foot living area are located the bedroom, kitchen, bath and a raised loft room. Across the way lies a four-hundred-square-foot section for a study and garage.

Wright likes to boast that his home has only two moving parts—the earth moving around the sun. And, in reality, his home is a hotbed of passive ideas. The building's orientation for maximum winter sunlight helps capture solar heat as well as make the inside bright year round. Wright used materials with large thermal mass everywhere he could in the house. The floors are brick. The rear walls are heavy concrete. Rooftop skylights allow the sun to penetrate to all parts of the house and can be opened to let out hot air on summer days. So successful was Wright's passive design in the mild California climate that the sun plus ten pieces of oakwood burned in a wood stove provided all the home heat during the winter of 1976–77.

Wright has also created a clever natural ventilation system in the house. At the bottom of his broad glass walls he has an awning window that can be opened on hot days. At the top of the walls is a pair of vents. The outer screen vent is always open, but the inner ceiling vent can be opened or closed as needed. On summer days he pulls insulating, radiating shades across the glass walls, keeping the hot sun out of the living area. However, the sun still heats the air sandwiched between the inner and outer panes of glass. As that air heats and rises, it sets up a flow that pulls air through the house without using one watt of power.

Another interesting variation in Wright's home is that it is comprised of far more wood than most earth shelters. Ceilings and interior walls are all striking blond slats that give the place far more of the feeling of a rustic cabin than an earth-covered building.

The unique two-section atrium pattern that he has established also provides for maximum view with minimum exposure to the elements by placing the smaller section between the main house and most of the wind and weather.

The interior of David Wright's sophisticated hillside cabin, which requires virtually no heating beyond the gentle warming of the sun.

BILL CHALEFF'S SUNTRAP III HOUSE

Last winter Long Island architect Bill Chaleff decided to practice what he preached and build himself an earth-sheltered home. Chaleff had already produced a pair of lightly bermed models for others; now he decided that he ought to do the same for his own family.

The result: Suntrap III, a below-grade, heavily passive solar design. The 114 × 20-foot house has 450 tons of earth on the roof, where Chaleff tends an impressive vegetable garden. The north concrete-block wall is bermed right to the roof, while the solar-collecting south wall is largely glass, with a heat-capturing greenhouse jutting out from one end.

"When I drive through the potato fields and along the dunes out here," Chaleff says, "I realize just how insensitive to the environment these New York architects are. They refuse to recognize that the visual landscape belongs to everyone who drives by. And aside from being energy wasters, these self-centered buildings are a visual blight on the landscape. The theory of architecture in the context of American life is absolutely bankrupt today because the leaders in the field refuse to take the responsibility to serve larger societal needs that transcend their own artistic goals and their clients' conceits."

Chaleff feels that the architectural "environment interacts more strongly than any other technology and it's up to architects to grapple with everyday problems and come up with solutions."

When he designed Suntrap III Chaleff had two principal goals in mind: to get as much passive solar gain as possible and to keep costs within the range of middle-class buyers. So he built the three-bedroom, one and one-half-bath test model for his future work of poured concrete and concrete block. He put quarry tile over the concrete of the kitchen floor to retain heat rising through the slab. The fireplace was made extra wide and reflective so as much of its heat as possible goes into the room, not up the chimney. Then the chimney was equipped with a heat exchanger to save some of that waste heat.

Chaleff kept the cost down by using as little insulation as possible and by choosing cheap pine walls instead of more expensive and common plaster or dry wall. The roof was built up from rigid construction foam placed over heavy timber beams instead of concrete. The foam was waterproofed with hot tar and four inches of stones above it to provide fast drainage.

Because he bermed so heavily for heat savings, Chaleff placed clerestory windows along the north wall for light and ventilation.

Doing much of the work—and all of the design—himself, Chaleff brought his home in at thirty dollars a square foot, 1980 prices. That's in one of the more expensive areas of the country. "Of course," he notes, "as an owner-builder I did a lot of the work myself and all of the contracting, which kept the cost down. But I just sent two similar houses out to bid and they came in at fifty-five dollars a square foot. Considering that conventional homes cost an average of sixty dollars a square foot out here, this isn't bad at all."

He feels it's conceivable to bring in an earth shelter for far below the cost of a conventional home if all amenities are foregone. "There are ways to save. My quarry tiles are a decorative addition. Poured concrete, tinted and scored for texture, would do the same thing. And I used a pattern on my fireplace wall—I call it poor man's stonework—also for decorative purposes that added about 5% on to the cost. Why one of the houses I designed recently for an owner-builder came in at fifteen dollars a square foot. Conversely, someone with lots of money could add even more elaborate features."

Chaleff uses a wood-coal furnace for backup heat and figures the system will gobble only seventy dollars for the heating season.

"There's an increased relationship to the outdoors" in an earth shelter, he believes. "You live in a solar house and yet become environmentally sensitive and thus socially sensitive. This fights alienation and builds a community. If all architects were forced to pay for the houses they design and then forced to live in them, there'd be a real revolution in architecture today."

CHUCK AND JANE ECKENRODE'S SUNEARTH HOUSE

In a small, professional family subdivision near Longmont, Colorado, stands the bermed up house of Chuck and Jane Eckenrode, easily the most unique structure in the complex. The Eckenrodes have lived in the house, designed and built by Colorado SunWorks, for about three years. The experience has had its good and bad points.

One aspect of the house the Eckenrodes like that others might find a drawback is the togetherness fostered by the design. "Because it is a passive solar home, it has a very open layout," Jane says. "This really necessitates a change in life-style, especially with younger children. You are always living with your kids. There are no escape areas and a lot of people wouldn't like that. We find it appealing and that's one reason we enjoy the house, besides the fact that it's comfortable and cheap to live in."

Cheap is right. Paul Shippee of SunWorks packed every passive solar and heat-saving idea he could come up with into the house. The three-bedroom eighteen-hundred-square-foot house has eight-inch-thick concrete walls and a foot of earth on the roof. Earth has been piled up against the north, east, and west walls, with the south-facing wall free for a host of passive gadgetry.

Twelve windows on the south wall soak up sunlight. Behind a large portion of them stand fifty-four fifty-five-gallon water-filled drums painted black. Water, as noted elsewhere in this book, has tremendous thermal mass, so these water-filled drums increase the ability of the house to store heat and release it slowly. At night bead-wall insulation is blown into the space between the double pane glass, sealing the day-gained heat inside. The warm barrels radiate heat directly into some rooms; their heat is carried elsewhere in the house by means of a clever convection system. Vents above the stack of barrels accept rising warm air, which circulates in a space between the walls to other parts of the house.

Ventilation patterns are designed for maximum summer cooling, as are the roof, with an overhang to block the summer sun, and the six skylights that line the north end of the house.

During the winter of 1978–79, this system provided 74% of the house's heating needs. Waste heat and occasional fireplace fires made up the rest. The backup gas furnace (a requirement of the mortgage company) was not used. The passive solar system also generated 25% of the house's hot water.

The prodigious energy benefits of the house were partially offset by social problems during the Eckenrodes' early occupancy. Their house was so different, and drew so many curious spectators, that their neighbors resented it. It took a long while before the neighbors accepted the family in the strange house. Now they find that the famous solar home makes it easier for them to give directions to their own places.

Another strange adjustment the Eckenrodes noticed was in their children's voices. They talked far too loudly at first in the new house. "They were used to talking with a lot of background noise," Jane states, "and it's dead quiet in here."

A pleasant side benefit for the family is improved health and better sleep because the environment is controlled. "Because of the tight construction level," Jane explains, "we maintain a 45%–50% humidity level, which is really good for the body. We have fewer colds and bronchial conditions. Cuts heal faster."

The only remaining question of consequence for the Eckenrodes about their experimental house is what they will get for it if they have to leave. "We do wonder about having problems with resale. That's a gamble we chose to take. We hope that as people learn more about these houses, resale won't really be a problem."

Any couple who bought an earth shelter until the past couple of years took a substantial gamble. The buildings were unproven; no one knew whether they would be salable in the years ahead. As more and more are built, though, the picture becomes ever clearer that earth-sheltered homes hold up well in the marketplace—better, in fact, than many surface dwellings. They don't wear out as fast and their energy-saving properties become more valuable each year.

Today the odds grow increasingly strong that the Eckenrodes won't have any problem selling their house at substantially more than they paid for it.

The SunEarth House's southern flank basks in the warming glow of the sun. Other walls are buried so that little of the solar heat is lost.

MALCOLM WELLS'S BERMED OFFICE

The site is a hill with a stunning view of Upper Mill Pond on Massachusetts' Cape Cod. It lies at the end of a winding dirt driveway, near enough to neighbors for convenience but with a sense of utter isolation, of sylvan peace and tranquillity.

In the midst of this forest scene stands Malcolm Wells's home and office. A long, narrow structure, it is nestled down into the soil up to the timbered roofline. Native wildflowers and grasses riot about on the roof, but the berm still hasn't had time to grow itself a strong ground cover. Wells believes that natural vegetation is the best possible cover for earth shelter soil. The native plants will withstand local temperatures and weather best, and they will be subject to natural selection, rather than lawnmowers and fertilizer, to maintain a healthy state. So Wells doesn't landscape his home; he lets nature do it for him.

The most immediately striking feature of the home is a huge skylight that covers the peak of the roof from end to end. The entire crown of the house is glass. In order to make the best use of his magnificent water view, Wells situated the house facing east-west. Then, in order to maintain some passive solar potential, he placed the triple-glazed skylight along the whole length of the roof. The result is an interior that is bathed in light. The light quality inside is softer but barely less intense than it is on the hill outside. Parts of the skylight open for ventilation.

To further increase the heating potential of the sun in his house, Wells placed "sniffer" ducts beneath the skylight. When the glassed-in air grows hot enough, these ducts draw it down to a maze of pipes embedded in a layer of sand that lies between an upper and lower concrete floor. As the hot air wafts through the pipes, it transfers its heat to the sand. This gets rid of excess heat in the summer, storing it in the ground, where it may take an edge off heating needs in the winter.

The home is heavily insulated with two layers of fiber glass glazing, wood shingles, six inches of insulation in the walls and twelve to eighteen inches of dirt all around. Yet on a hot summer day it doesn't feel at all closed in. In fact, the sense a visitor gets is of a slightly enclosed patio. The structure doesn't seem to be enclosed by the earth; rather, it seems to be part of the earth, a natural adjunct to the surroundings.

This is the feeling Wells was after when he designed the home. As he has stated, "Once we step across the line and begin to sense the magnitude of this outdoor miracle with which we so casually tamper, architecture will surely change for the better."

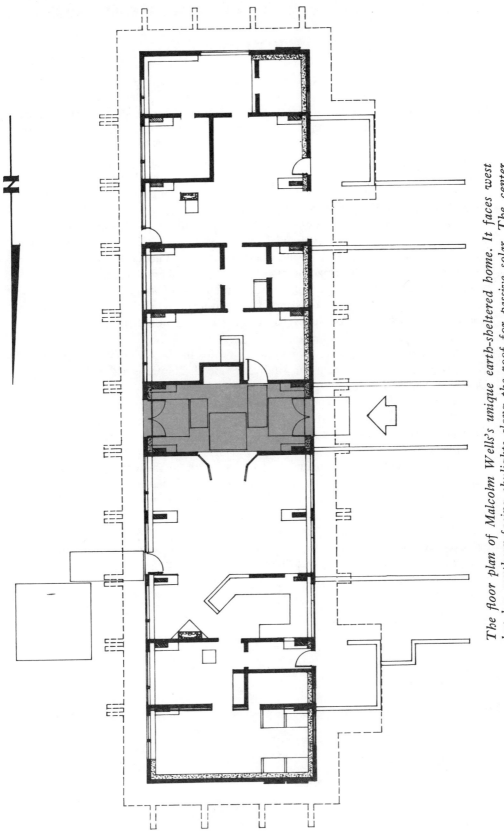

The floor plan of Malcolm Wells's unique earth-sheltered home. It faces west but has southern-facing skylights along the roof for passive solar. The center section is a skylit atrium that links the two halves of the house.

JOHNSON'S LAKESIDE ROMANCE

Philip Johnson was one of the first architects to see the possibilities for modern earth-sheltered design. Way back in 1965 he placed an opulent suburban home, the Geier house, in the crook of a lakeside hill near Cincinnati. His motivation, he says, was "romantic reasons," but his design, while stunning, was also quite practical.

Given the year in which it was built, Johnson's mound home was quite a breakthrough. Until that time virtually all underground design had been predicated on notions of security; beauty had been considered an option available only at the surface. Johnson changed that.

Johnson was inspired, he says, by the beauty of the site his clients provided for him. "It was a beautiful field and I felt its connection with the water was so important that why not keep it just the way it was. The Geiers had ponies at the time and they could graze on the roof. The dwelling affords complete privacy from the main road, and with water coming onto the house as it does, you have the feeling of being on an island, away from the world, although you're in the middle of suburbia. All the Geiers can see looking out their window is their own land rising on the other bank."

Rather than facing an atrium or vista, all rooms in the Geier house have connection with the water. From the landward side the house is virtually invisible. Only the chimneys give any sign that people live here.

Johnson, who loves "the cuddly feeling cellars give you," set the Geier house beneath fifteen inches of earth. The walls consist of a foot of poured concrete, the roof six inches. The entire structure is waterproofed extensively, he says, "just like a boat."

Johnson's triumph in the Geier house is that, with all the structural underpinning and insulation, the building looks effortless and so natural it almost escapes notice. The best earth shelters are those in which the solutions to the structural problems actually make the building more handsome or more enjoyable.

Here, faced with a tricky site, Johnson crafted a building that uses the hill overhang for privacy, the water for beauty and an expansive feeling, and the required insulation to keep everything quiet so that the owners would never have any trouble hearing the gentle lapping of waves against the front porch.

THE PENDELL HANDMADE HOME

When Claude and Virginia Pendell of Spokane, Washington, were looking around for a retirement-home design, they lit on earth sheltering for several reasons. They wanted a small place that wouldn't take much work and wouldn't eat up much of their money. And they wanted a home that would be simple.

Claude, a horse trainer with some experience at building, put the home up pretty much by himself. He created a simple boxlike one thousand-square-foot design that would be buried in a hill with both the south and west walls exposed. He placed as many windows as he could along those two surfaces so that the quality of light in this house would not be too different from what it was in the couple's old, many-windowed surface home.

To a nonprofessional builder the roof represents the major challenge and danger. Claude spent most of his time on the roof's design and construction. It slopes downward slightly toward the rear, with six to eight inches of dirt on top. The support for the soil load consists of rough-hewn two-by-fours spiked together face to face, resting upon 4 × 12-inch laminated beams. The roofing platform is supported by thick log posts and strategically placed bearing walls.

Claude covered the outside of his roof with plywood to protect the plastic liner he used for waterproofing material. Contrary to accepted practice, Pendell put no waterproofing at all on the walls because he felt that in this area the soil would draw moisture out of the house instead of vice versa. So far he seems to have been proven correct.

Similarly, he decided to rely entirely on the temperature averaging of the soil to keep his house temperature balanced. He put no insulation in the house at all. If it gets too hot inside, he says, they'll just open a door and let some fresh air in.

The flooring, too, was extremely simple. Eight to ten inches of concrete was poured over a bed of rough gravel. Wood slats hold up the plywood subfloor, covered with a thin layer of particle board. On the south side the floor was held back six inches from the house edge so the sun can hit and warm the underlying gravel. A tiny squirrel cage fan at the north end of the house draws air over the stones so it can be warmed—a down home passive solar collector.

The Pendell house represents low-tech architecture at its best: small, simple and not overly ambitious. They know their home isn't an architectural showpiece, but that's not what they wanted. They like it, and they're surprised that anybody thinks it's something to make a fuss over. "We're very happy with our home," Pendell says, "but it's an insignificant house and we're embarrassed because it's so small yet so many people want to see it."

The Pendells are proof that amateurs can build their own earth shelters provided they keep everything simple and stay near the surface so the weights and tolerances don't get out of hand. But keep in mind that Pendell was no suburban greenhorn. He had a mechanical background and prepared for his building carefully. And he was willing to make sacrifices in size and fanciness to keep the project within the scope of his abilities. He knew he couldn't have it all, so he didn't try to get it. He settled for what he could manage.

The Pendells built a basic, homemade earth shelter that might be too elemental and too small for many but is just fine with them.

THE LIERLY'S CAVERN

Price and Sylvia Lierly live in one of Jay Swayze's patented Geobuilding homes. Their entire living environment, both "interior" and "exterior," lies in a subterranean cavern. In this most underground of earth-sheltered situations, the surface isn't something you interact with but a separate place to go to when you want groceries or a restaurant meal.

Here is how one of Geobuilding's own pamphlets describes the enclosed environment they create:

As you approach the building you see a beauty-wall which serves as a dirt retainer as well as providing an eye-pleasing sight. The entry is through the beauty-wall. One or more decorator-styled exhaust structures are pleasantly evident, along with the striking landscaping.

A wide, well-lighted entry leads you into the controlled environment. First, you are greeted by a spacious courtyard complete with arbor, flowers, and natural sunlight in the yard. Clean, fresh air combines with the quiet atmosphere to make you aware of a building as you have never known before.

For families the Geobuilt home offers an underground exterior and interior. As you leave the courtyard, you step onto a porch, go through a door into the interior of the home. Two or three bedrooms, two baths, family room with cathedral ceiling and fireplace in the family room. The effect is the same on everyone—you simply forget that you are below an earth covering. The interior is the same as conventional luxury homes, plus the advantages of a controlled environment.

As you look through a window in one of the rooms, you view a beautiful land- or seascape painting. And, yes, you even feel a light breeze from the natural flow of fresh filtered air caused by the patented design of the building.

The Lierly's place looks from the street like a typical Spanish-style ranch. But that is just a facade. All that actually stands on the surface is a two-car garage and Sylvia Lierly's tax office. Everything else—the living area, the two bedrooms and the yard—lies below grade.

The yard has a grasslike lawn and a large, ornate skylight to let in some sunlight. Elsewhere in the twenty-two-hundred-square-foot home a sense of surface living is provided by optic lights, which can mimic the passing of the sun overhead, and by Swayze's patented air filtration and circulation system, which creates variable breezes.

Does living in a concrete cavern make the Lierlys claustrophobic? "Oh, no" says Sylvia. "It doesn't feel that way at all. We have the solar light in the courtyard and windows in the house looking out on the courtyard (which, keep in mind, is underground). We get light through a solar bubble. You don't feel like you're underground at all."

The Lierlys chose a completely enclosed home because they felt it would

bring them the maximum energy savings and security. They point out that their
tiny amount of contact with the surface allows for total environmental control:
no heat leaks, no compromises on energy efficiency. They feel comfortable, too,
with the fact that their home is virtually burglarproof. "You like to leave society behind in your house," observes Sylvia. "You don't want to have to worry
about what's going on outside or who might try to get in. Now we don't."

All told, the Lierlys are overjoyed with their subterranean minisuburb. They
think it could easily be the wave of the future. "I really think it could," Sylvia
states. "I think when people see houses that look like this and feel like this,
they're going to go for it."

*The yard, as well as the house, at Price and Sylvia Lierly's homestead lies en-
tirely underground.*

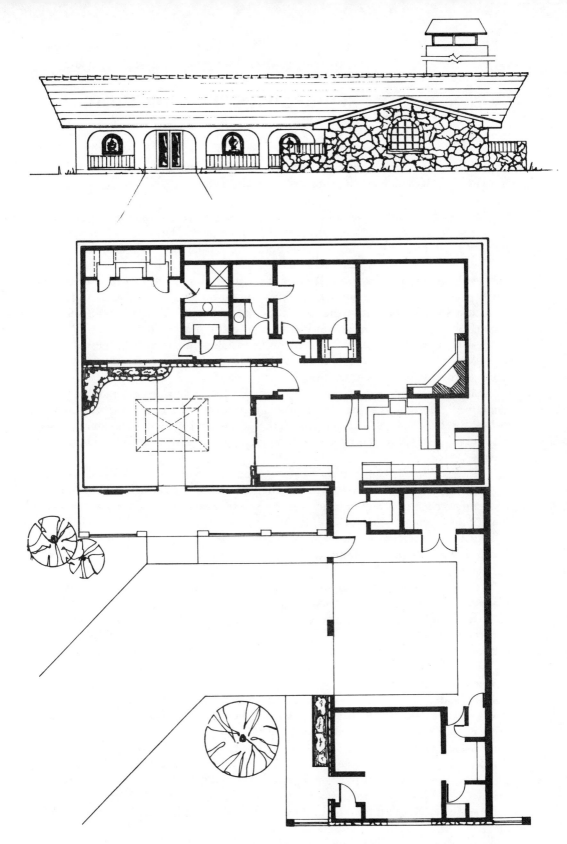

From the street the Lierly home looks like any suburban mansion, but it's just a facade. The real home is a man-made cavern with its own underground yard. The floor plan shows how extensive the home is. The rectangle with an X at its center, located just to the left of center, is a skylight over the subterranean patio.

RON AND RUTH SCHROEDER'S HILLSIDE HOME

Ron and Ruth Schroeder have a home in exurban Fairborn, Ohio, that is burrowed snugly into the hillside of their five-acre lot. The seventeen-hundred-square-foot two-story dwelling fulfills the couple's desires to live a more self-contained, self-sufficient life yet still remain comfortable—and just a touch elegant.

"Our house is more open and airy than the house we had in the city," Ruth says, and that's not surprising, because the foyer of the house stands beneath a twenty-four-foot skylit ceiling that the Schroeder's call their "solar atrium."

The couple use the atrium as a passive solar collector. It stands behind the surface-level south-facing wall. As in other passive designs, this wall is made of thick glass that traps the sun's heat behind it during the day. This heat forms a thick cushion of hot air at the top of the tall atrium room. A fan pulls this air out and downward into the rest of the house, thereby heating it.

Heat-retaining brick walls in the living and dining rooms soak up heat during the day and slowly release it at night, equalizing house temperatures and increasing the effectiveness of the passive design.

At the Schroeders' the atrium and all of the roof stand above grade and are unbermed. Most of the living area, though, stands well back into the hillside below the frost line. Soil temperature averages around 55° all year round. The walls are composed of ten inches of concrete covered with rigid foam insulation board.

The reason for leaving the roof bare was purely economic. The extra weight of earth on the roof would have required so much more structural support and concrete—some ten thousand dollars' worth—that "it would have taken a long time of using less fuel to save that much money," according to Ron.

When they decided to try earth shelter, the Schroeders selected a professional underground design firm for the plans and a custom home builder they knew personally to build the house.

Even with professional support, though, they found financing tricky. Many places they tried simply told them to get lost. "A lot of the places we talked to said we shouldn't even bother to apply because they didn't make loans for underground houses," Ruth recalls. To avoid disappointment the pair checked out the loan policies of banks before making applications. When they finally found a potentially receptive ear, they made every effort to blow away all possible objections to their design.

They added an electric furnace to the plans, even though they felt their wood stove would be all that was needed. They hoped it would convince the bank of the home's resale potential. Then Ron, a commercial artist, drew a color sketch of the home to show how good it would look. They also made a detailed pitch about how the reduced burden of energy payments would increase their ability to meet mortgage payments without strain.

It worked. They got the loan and built the house—proof that careful planning and a willingness to compromise can overcome the prejudices of loan officers against earth shelter. If the Schroeders had been inflexible or ill-prepared, they might still be living in their old house.

ANDY DAVIS'S CAVE

Andy Davis lives in Armington, Illinois, a town of 350 people located roughly in the middle of the triangle formed by Peoria, Bloomington and Springfield. His home is right on the edge of the village, a cornfield coming right up to his lawn.

At first glimpse, the house looks like a Hobbit hole—with its round windows, oblong door and stone wall—set in a low mound. The front door knob is an animal bone. The interior walls and ceiling are all stone cast in concrete. Potted plants sit in every nook and cover the walls—testimony to Davis's claims about the sunniness of his cave home. Synthetic animal fur covers the front door and many of the other doors, giving the home an almost disorienting sense of newness.

Davis himself looks like the typical small-town farmer or mechanic. He's about six feet tall, slightly paunchy, slightly stooped. He has a Will Rogers grin and once in a while he will grimace as he talks. He wears jeans and a blue cowboy shirt.

"I was born in the area and have lived in small towns around here all my life," he says. Until a few years ago he was an electrician, but his life began changing in the summer of 1973 when he visited relatives in Arkansas. "We were poking around in the fields one day, and I went back into an old abandoned mine shaft," he said. "It was over 100° that day. There was a lot of dampness at first. But as I came back in the cave it began to dry up. Then way back it was perfectly dry." Davis also noted that "the rock walls were very pretty." He sat down to rest and thought that if it was cool in summer it must be warm in winter.

In the autumn of 1975 Davis sold his farmhouse and "rented a house in Armington. There were no storm windows, and the curtains would blow back and forth." The first monthly heating bill was $167, and this in a house so cold that "if you wanted to sit and watch TV, you had to sit with a blanket in your lap."

He decided to build a home. "But I just couldn't bring myself to build a conventional home," he says. Ever since the visit to the mine shaft, he had thought about the possibilities of cave living as an energy solution.

First he had to convince his family. "I didn't push it, just kept talking and explaining." He won his wife over, but his children still were leery of his plan; they feared their schoolmates would call them "The Flintstones." Finally the children saw their father was going ahead anyway, so they pitched in. The whole building was built by Davis, relatives and friends.

He found a cheap site on a low rise, not quite as steep as he would have wished. He dug a semicircle and poured the concrete footings, walls, roof and floor.

He then covered it all with four feet of dirt and landscaped it. "I wanted to simulate what it was like twenty thousand years ago," he says.

The two main problems he faced were dampness and the need to support the weight of the earthen roof. The dampness turned out to be little trouble. And his reinforced concrete roof has supported trucks filled with dirt, construction vehicles—even a runaway herd of cattle.

Inside, a large living room and three bedrooms all face the front door and the main windows. "I can see out from every room," Davis says. "I had claustrophobia, and I wanted to be able to see out at all times."

A small kitchen area is set to one side of the living room. The three bedrooms were rather small but attractive. He has the synthetic animal fur on the doors and some of the furniture. A bone holds a roll of toilet paper in the bathroom; nooks in the rock walls serve as shelves.

He admits to three mistakes. First, he gave the kitchen cabinets a rocky cave look, but they are hard to clean. The rocky ceiling was too dark, and the recessed lights in the ceiling were too inefficient. The artificial fur, though, was easy to clean. "As long as I was at it, I thought I'd make it unique," he chuckles.

As for his claustrophobia, Davis found his cave home helped him get over it. "The first night I slept in here I never slept so good in my life," he said. Now he and his wife find they cannot sleep well when travelling.

Andy Davis says that the only drawback of his earth-sheltered home is getting out on weekends to mow the roof.

"Everyone who gets one, after living in it awhile, gets to feeling insecure in a house above the ground," Davis says. "Here you're protected from fires, storms, rain, snow, tornadoes, hurricanes and everything else." He has no fire insurance.

The round windows in front are the only ones, but the home is fairly sunny. The home is also warm, thanks to a Franklin stove in the living room. The Davises used two and one-half self-cut cords of wood the winter of 1976–77, and three cords last winter—the worst and coldest on record in central Illinois. Fuel costs were $1.29 the first winter and $4.19 the second. Last year, according to Davis, "my son broke an ax handle [and] I had to buy a new one," thus explaining the cost increase.

The cost of Davis's home was about fifteen thousand dollars, but it is impossible to calculate how much a Davis franchise house will cost because much of the work on the original home was done by the family, and the plans were Davis's own. However, Davis notes that the franchises he operates price their homes about 10% below the prices of other homes in those areas.

Looking ahead, Davis hopes to establish a self-sufficient village of cave homes, which is now in the planning stage. It calls for twenty-nine cave homes to be built in Armington. The community would have a small airstrip for the visitors Davis expects at the pilot community.

The key to it is a large hog house. The methane produced by the pig feces would provide heat for the whole community. Davis has just begun talking to federal and local officials about the plan and says he has buyers eager to buy the homes in the proposed communities. The cave hog houses are already a great success, a harbinger of projects Davis hopes to work up. Hogs raised in one of Davis's underground hog houses need 20% less food, Davis says. Rats and bugs can't get in, so there is less disease; and snug under the earth, the hogs suffer less distractions from noise, weather, and so forth. "They're just more contented."

Davis foresees underground churches and office buildings as being more economical to heat and light, safer, and using space more economically. "People only moved out of caves when their water ran out, so they went and piled up some sticks and kept on going that way," Davis says. "If people had had a way of ventilating caves and a way of making new ones, they never would have moved out."

PART FOUR

COMMERCIAL
EARTH SHELTERS

If earth shelter were merely an architectural form that applied to single-family development, it could be labeled a curiosity. But it is more. It is, first of all, an exceptional method for developing suburban communities, an ecological version of the brick ranch house boom of the fifties. "The underground concept," states John Barnard, "is eminently practical for suburban communities—single-family homes, apartments and commercial buildings. Such communities would resemble parks rather than the urban sprawl that blights much of the country."

But that is not all. Earth shelter also represents a basic, important form of commercial development. Factories, shopping centers, offices and schools are all moving beneath the protective blanket of soil.

In reality, use of earth-sheltered space has a long history in the commercial sphere. The mall beneath New York's Rockefeller Center is half a century old. Les Halles in Paris sat atop a warren of subterranean working space. Safe City in the Hudson River valley was carved into the rock slope of a cliff to provide ultimate storage and living quarters in the face of increasing violence and nuclear peril. Then, of course, we have subways, tunnels and dozens of other civic uses for areas below the surface.

MILITARY USES

Most of all, up to the present day, we have had the military, which, like a dog, loves to bury its bones. Deep beneath Cheyenne Mountain, near Colorado Springs, Colorado, for instance, the North American Air Defense Command Combat Operations center comprises 11 3-story buildings with a total floor space of 176,000 square feet all under the surface. The complex contains 3 parallel chambers, each 45 feet wide, 600 feet long and 60 feet deep. The buildings are mounted on springs in case of vibrations from earthquakes or enemy attack.

Similarly overwhelming is the Swedish government's naval base at Muskö Island. Caverns large enough to comfortably hold warships have been hewn out of the rocky island. Gigantic fans haul off the exhaust fumes so the ships can stay under power while under cover. Air locks are designed to keep out radiation—if it comes to that. Sweden has also dug out underground space as hangars for 1,500 military jets, aircraft plants and a 1,600-bed hospital.

Back home we have a couple dozen subterranean missile silos scattered about the countryside, each 174 feet deep and 52 feet in diameter, about the size of a round 17-story building. And in Greenland America's military force has put together Camp Century, which contains the first prefabricated, portable nuclear power plant designed to handle all the needs of a self-sufficient community. The buildings—4 levels of them—are made of paper and resin mixed with frozen snow and ice. Some 250 people involved in polar research live and work in this frozen labyrinth.

And, lastly, there is America's nuclear command center burrowed into Thunder Mountain outside of Washington. There our government will stash vital documents if Armageddon comes, supposedly protected from the nuclear ravages above.

The military's reasons for going underground certainly aren't those of home or business owners, but the armed services have shown that large-scale earth-sheltered development is possible. In the past few years commercial developers, influenced by the fresh ideas of the new underground-housing architects, have begun melding public functions and the soil.

COMMERCIAL USES

Forward-looking designers have applied earth shelter principles to buildings ranging from a church for monks in Benet Lake, Wisconsin, to Cleveland's massive convention center. They have found that the use of earth, far from limiting the effectiveness of commercial buildings, creates unique situations for functional beauty, as well as offering other advantages.

Philip Johnson, one of the earliest earth shelter dabblers, sank a small art gallery beneath grade on his estate in Connecticut and found that the underground location protected the artwork better from harmful sunlight and dust. Moreover, the even temperature of the earth shelter meant less wear and longer life for priceless works.

Businesses and utilities have found uses for earth sheltering, too. Sweden, France and Norway have sunk nuclear power plants beneath thick earth collars to provide maximum safety and security. In the United States, companies have turned downward for storage, manufacturing and office space. They find the environment below deck to be cool and controllable, noise and dust-free, and not a bit unsettling for workers.

At the Wampum Mine Storage Company, for example, America's civil defense medical supplies are housed in mined-out limestone caverns. An official of

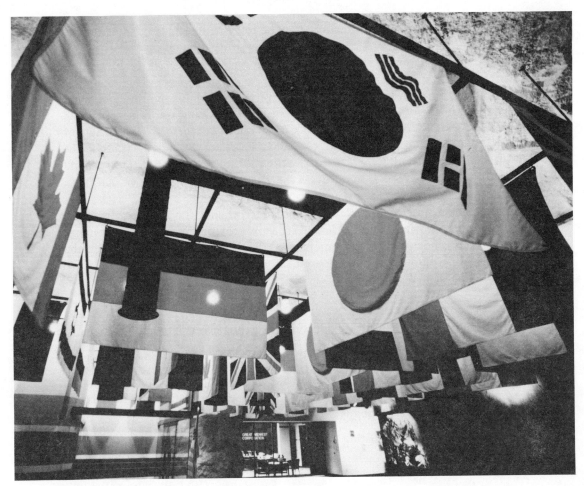

Flags bedeck the natural limestone ceiling of the caverns beneath Kansas City's industrial park, where dozens of businesses have set up shop in recent years.

the firm states that "comparative costs show that, even taking into consideration mining expenses, setting up underground warehouses to provide maximum storage benefits is economical."

The grandest commercial subsurface development to date, though, is certainly the underground industrial park at Kansas City, Missouri. Here, in six square miles of leftover limestone mines that range from forty to seventy-five feet below grade, some two thousand people work each day. Two dozen businesses house themselves in the caverns for a wide variety of reasons.

The first operation to go below ground in Kansas City was Brunson Instruments, an optical firm that made the surveying equipment used on the moon. In order to precision-grind their lenses, they wanted to be free from street vibrations and noise. They also wanted a place where dust could be carefully controlled. They found it in the caves.

The complex's largest tenant is Inland Cold Storage Company. Because the company's refrigerators never have to fight off surface hot spells, they can provide refrigerated storage more economically.

In Kansas City it has turned out that leasing the old mining space is more profitable than the original mining ever was. The increasing use of subterranean space has increased the size of Kansas City without recourse to annexation or energy-wasting skyscraper building programs. Professor Robert Gentile, of the University of Missouri–Kansas City, reported that mining operations around Kansas City add twenty-five acres a year to the city's rentable subterranean space.

As the first deliberate commercial move underground, the Brunson operation has been rather heavily studied. The results show that the firm netted even more benefits from underground adaptation than they'd planned on. They got the vibration-free environment they desired. In addition, according to a study by Professors Bligh and Hamburger, the latter of the University of Minnesota, they

- saved maintenance due to lack of wind, moisture, excess heat and freezing.

- received lower insurance rates because of fireproof environment and protection from wind damage.

- got breaks on installation costs of utility lines; the latter can be placed more freely since freezing is not a consideration.

- saved money by eliminating costly shoring for heavy machinery; the earth made an extra support unnecessary.

- reduced the need to shield sensitive equipment from outside vibration; since there wasn't any, there was little need for the shields.

- derived savings because machines didn't need to be realigned as often; expansion, contraction and humidity changes were reduced, lowering machine shifting.

Here is Bligh and Hamburger's comparison of the energy and cost requirements of Brunson as compared with a similar operation at street level:

ITEM COMPARED	ABOVE GROUND (ESTIMATE)	UNDERGROUND (BRUNSON CO.)
Heating Units (BTUs per hour)	2,000,000	750,000
Refrigeration (size of unit required in tons)	500–700	57
Dehumidification (size of unit required in tons)	50–70,000	3,200
Fire Insurance ($/$1000)	2.85	0.10

Another example of significant underground commercial development in the Middle West comes from Mutual of Omaha, a staid insurance company. In 1978 this cautious, conservative corporation began work on a new downtown office building, three stories down, with a huge glass dome at street level.

The reasons for the turn downward were clear and businesslike. According to William L. Enenbach, executive vice-president for administration, "First and foremost among these reasons was that underground construction allowed the companies the most efficient utilization of the land that was available. The only available site with a close proximity to the home office was adjacent to the north side of the building, in the form of a parking lot. The home office, however, stands twelve stories high on the north side, with a fourteen-story tower connected to the center of the main office structure to provide additional space. Company officials wanted to preserve the appearance of the building on the north side but realized this was not possible if an aboveground building was to be constructed. The ideal solution, therefore, was to build underground."

Energy conservation also figured in Mutual's decision. Their research found that below the frost line Omaha soil holds a fairly constant 50° temperature. A new underground office would only have to generate 20° of heat to keep employees comfortable. When Mutual figured the heat given off by the workers themselves, the lights, machinery and kitchen facilities, they found that the gap to be bridged by generated heat was small indeed. The earth's heat sink would flatten out their energy needs. A building with no windows would be inherently more energy-efficient. So Mutual went down.

The company was overjoyed to find that the cost of building the office space in the earth was far less than it would have been on the surface. The building's estimated cost was twelve and one half million dollars, about five million dollars less, by company calculations, than the same structure on the surface.

Not every commercial earth shelter will be so fortunate. Mutual's case offered some special benefits. In large part, the cost savings can be traced to the cost of the building materials. With an underground structure, the four exterior walls consisted of poured concrete. An aboveground building would have necessitated matching the finish of the existing home office.

The company's final reason for going down was that it would disrupt parking less than a new surface building.

Despite the enthusiasm for an earth-sheltered office, Mutual did find some drawbacks. The excavation, for instance, took much longer than for a surface building. Mutual's hole required the removal of 126,000 cubic yards of dirt, which took nearly nine months of intermittent work. They also found that rainy weather gives underground builders even worse fits than surface ones. Finally, the expense of waterproofing was something of a damper.

In the end, though, Mutual took the plunge and the resulting building is one of the prides of Omaha. The bronzed aluminum dome, with its surrounding garden courtyard, brings some much-needed street-level spunk to a drab downtown area.

Mutual of Omaha is not alone in thinking about the earth as the best place to develop new additions to high-density urban areas. It is a way of making new space available without disturbing the sight or view of the existing skyline. There is a deeper question of whether a commercial development requires, or even deserves, a spot in our visual landscape—without any other need to be there.

"There are too many individual buildings today," wrote Gunnar Bickerts in a report entitled *Subterranean Urban Systems*. "Not every physical or functional need deserves the right to become a visual object in our landscape. Nor does it have the right to occupy a piece of land, exerting its visual effects. Most likely its presence is not needed for the formation of our urban fabric. We have to impose a birth control upon certain buildings and other structures in order to check the ugliness of urban sprawl."

Arthur Drexler, director of the Museum of Modern Art's architecture and design department, believes that an excellent way to achieve that end would be to use subsurface design: "An awful lot of things that have to be built don't require or merit architectural treatment, in the sense of being thrust forward into your consciousness as statements about material or space or anything else; they have no particular intrinsic interest. Architecture is still thought to be a matter of buildings when it ought to be something else. Today all of our buildings are designed as large, useful objects. Each year we put up thousands of warehouses and factories, for example, which have no business existing as objects at all. They are services, means to an end. Why are they not concealed? Services belong in the ground. We should insist that whatever services are required be invisible, not beautiful."

The point is not that everything should be buried but that those things that offer nothing visually and can function well below ground might as well go there. "No one recommends burying everything beneath the surface," states Ken Labs. "It is neither feasible nor desirable. Nevertheless, a host of functions exist which are, by necessity, already well suited for and able to benefit from underground location."

Labs lists four groups that might be well served by earth-sheltered development:

INDUSTRIAL/CIVIC WORKS	COMMERCIAL/INSTITUTIONAL
waste-treatment facilities	*department stores*
transportation systems	*supermarkets*
telephone exchanges	*shopping malls*
power substations	*restaurants*
light industry	*nightclubs*
assembly plants	*theaters, cinemas*
warehousing, bulk storage	*concert halls*
refrigeration, cold storage	*museums, galleries*
parking lots, garages	*convention centers*

PRIVATE SERVICES	PUBLIC SERVICES
bus terminals	*libraries*
repair stations	*laboratories*
transport depots	*schools*
hotels, motels	*research centers*
recording studios	*housing, offices*

The American Society of Civil Engineers found that such development could unlock important benefits for society as a whole:

"At least three or four opportunities provide high potential for improvement when the underground frontier is examined closely. First, it will release *resources* or their surrogate *money* to apply elsewhere than where money is now applied in our systems. Most of the applications are presently constrained to the surface. Second, the new [underground] frontier can free *space*, which grows dearer as the urban areas expand and become congested. Third, underground development offers increased *flexibility* and expands the numbers of options in efforts to strike new balances in the urbanization process. Fourth, strategic development and use of underground space afford savings in one of the dearest commodities man possesses, namely, *time*. These factors (resources, space, flexibility, and time) present both an opportunity and a challenge to man's creativity."

EDUCATIONAL USES

This challenge is being met on many fronts. At the moment schools seem to be in the forefront of trying earth-sheltered designs. Perhaps the heavy financial strains on schools today have pushed educators a little faster toward new alternatives that can save them energy dollars. They are faced with dwindling budgets and decaying buildings. When a new building is required, they have to listen seriously to any idea that won't cost much more to build and will cost far less to operate.

In Oklahoma twenty-seven schools are at least partially underground and fifteen more have bermed walls. Originally the state considered underground space as a good way to get student populations away from the vicious twisters that snaked around the plains each year. But administrators there have found many other benefits in using the earth as an aid to education:

Preliminary studies by the staff at the state department of education have shown that those schools perform exceptionally well as learning environments. Both teachers and principals alike have commented on the lack of noise and distractions and the ease with which they could hold the attention of their students.

- There are indications that revenue requirements for energy and maintenance of underground schools are likely to be significantly less than requirements for comparable aboveground schools.

- There are possibilities of making dual use of available land by building underground.

- There are possibilities for some savings in utilization. Landscaping for aboveground buildings usually involves more expense to esthetically balance the view.

- The underground design also minimizes vandalism problems in schools by limiting exposed walls.

- Teachers can make better use of classroom walls to stimulate learning, since there are no windows.

- Lack of dust in underground schools provides a positive psychological effect in the relief of chronic allergy symptoms.

- There is a sense of security felt by community members in knowing that they (in addition to the children) have a shelter from tornadoes.

The Oklahoma experience did show that construction costs for the schools were "slightly higher than comparable surface structures." But administrators felt that "while the initial costs are higher . . . these higher costs are offset by long-term savings," according to a state report.

In Reston, Virginia, another school, Terraset, has become a tremendous success—happy students, happy parents and so many visitors that a reception center has been set up apart from the school building itself. The school has a subsurface central courtyard flanked by a multipurpose hall. Four rounded modules comprise the teaching area, which is linked to a skylit media center. It is quiet, efficient and beautiful.

But getting the school built took a Herculean effort on the part of the architects. They had to overcome the skepticism of, in turn, the school board's design and construction department, the state education authority, local building officers, fire marshals and the community. In the process, significant changes in the design took place to offset building inspection demands and bureaucratic worries. Occasionally the architects had to pull their way past pure prejudice. In the opinion of Douglas Carter, the chief architect, the state education authority "was very negative, to say the least. One of the energy-conserving items we'd come up with was the fact that a circle will enclose the same area as a square or rectangle with a much smaller peripheral surface area. Obviously, the smaller you make your wall surface, the less energy you're going to lose through it. That was the origin of our circular learning centers, and the state education authority thought we were nuts."

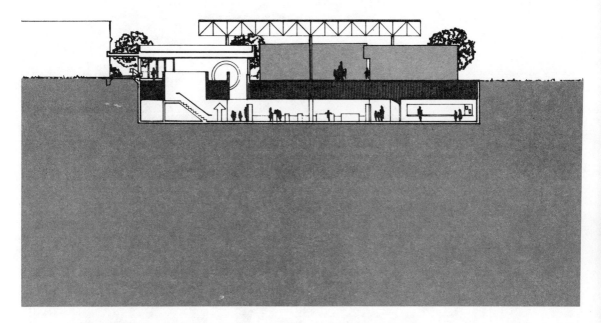

Terraset, the earth-sheltered school in Reston, Virginia, has proven so successful and popular, that it has become a prototype.

Still, the concept prevailed, largely because the community and the local officials came to feel that the operational savings and environmental benefits were worth the trouble.

In a similar situation, architects in Fort Worth, Texas, had to sweat blood to convince a Mexican-American community that an earth-sheltered school would be the best design for their small school site at the end of a jetport runway. In the face of confused, intractable community opposition, the architects trotted out examples of other successful earth-sheltered developments. They noted that these schools were quiet and hadn't needed paint for years. Finally, they pointed out that the Fort Worth library had recently gone underground. "The fact that the city would put its library underground," says architect Parker Croston, "meant to these neighborhood people that below-grade buildings can't be second class. That kind of thinking made all the difference."

Colleges are going down into the ground as well. The University of Minnesota took the first step, which is fitting, since it's home to the American Under-

When the University of Minnesota submerged Williamson Hall, the student center, energy costs plummeted and services didn't suffer one iota. More university buildings are currently beaded under the soil.

ground Space Association. About four years ago the university asked architect David Bennett to design a new campus bookstore and student center that would preserve the view of two historic campus buildings and wouldn't block the pedestrian flow in the crucial campus area. The building also had to be highly energy-efficient.

So Bennett took Williamson Hall 95% below grade. He put a clerestory window above the bookstore for light and placed an interior courtyard under glass angled at 45° so the lower floor could receive some light. He situated large planters above the courtyard to screen the summer sun but let in the needed winter light. The building has become an angled, light-filled showplace on campus. And it cost between 3% and 5% less than a similar surface structure would have, while using about 45% as much energy. New solar collectors should decrease energy use still further.

Williamson's performance pleased university officials so much that they've decided to try earth shelter again. With $15.8 million from the earth shelter-supportive state legislature, Minnesota is sinking its civil and mineral engineering departments into an ultramodern earth-sheltered office building. In addition to earth buffering, the 150,000-square-foot facility will use solar panels for heating and a clever system for storing winter heat to use in summer cooling.

At the opposite end of the country, the University of Florida dropped its Florida State Museum into a hillside. The bermed building rests on three terrace levels under sophisticated concrete sun canopies that shield against the summer sun. J. C. Dickinson, Jr., the museum's director, has called the earth-sheltered design "the most economical building on the entire campus. The berming and burying the building into the hillside is the answer, and right now the university is constructing a huge auditorium that is buried a story-and-a-half underground. I think that's the way we'll all be building in the future." During the energy crisis of 1974, electrical use in the museum was cut back drastically, including the complete shutdown after 5 P.M. of the environmental control system for the artwork. The temperature in the carefully designed earth-covered building varied hardly a whit. The precious artifacts were placed into the hillside, where the constant earth temperature of 73° kept them safe and cozy at all times. Offices and workrooms were placed facing a central courtyard, but even these more exposed rooms remained relatively comfortable.

Governments have jumped on the bandwagon, too. In Australia the new national parliament building is being recessed into the top of famous Capital Hill. The design of the site, by architect Romaldo Giurfola, creates an architectural imprint that expresses the relationship between nature and construction and, by extension, between the imposition of government power and the natural state from which the government evolved.

Rather than a massive building on the hilltop that would dominate the skyline, the architect chose to mold the building to the surroundings. For years the hill has been dominated by a single flagpole flying the Australian flag. The parliament design preserves the symbol of the flag in a tall flagpole set on a slim, four-legged tower. It is the only part of the building that will impinge upon the view.

William Morgan's bold design for the Florida State Museum proved to be comfy for workers, fascinating for visitors, and cheap to maintain.

Australia's new parliament building will cuddle under the crest of a hill near Sydney. The national flag rather than the building will dominate the view.

The rest of the parliament structure consists of sweeping curved sections that arch along the shoulder of the hill. Within this submerged circle a line of formal meeting rooms stand between two curvilinear walls that frame the chambers of the Senate and House of Representatives. From the air the design looks something like a huge baseball buried in the soil. It will almost surely become a classic of modern architecture.

LARGE-SCALE HOUSING

Beyond strictly commercial potential, earth shelter also holds great promise for large-scale housing development. The first homeowners to go below the soil were pure individualists, isolated and uncoordinated with each other. But now earth shelter has grown to the point where professional developers are getting involved. They may bring an explosion of underground housing.

Where individuals have had problems finding loans and qualified contractors, big-money developers have the clout to get what they want from building suppliers and financiers. Their willingness to undertake a large-scale project carries with it a bureaucratic weight that no group of individual homeowners could hope to achieve. The smell of a developer's big money will stiffen a bank's backbone and send an eager contractor back to school for special instruction.

The commercial potential isn't all on the drawing boards, either. Frank Moreland has recently been at work developing a subdivision near Dallas—six hundred acres of homes, all underground. He points out that an earth-sheltered development can leave up to 96% of the land in greenery that will provide gardens and fresh air. A typical suburb has only 56% or less of green space available.

Moreland and Thomas Bligh both foresee significant development of earth-sheltered subdivisions containing six to eight houses per acre. That would provide a saving to buyers because the homes would be about 20% more dense than in a surface subdivision. But careful site placement would eliminate visual pollution; it would be possible for each home to see nothing but trees. And the quiet of earth shelter would eliminate noises from neighbors.

Moreland also predicts the development of town house projects in which one floor's earth-sheltered roof would be the next floor's front yard—built at far less cost than today's homes. Within three to five years we will probably have underground houses of this type which can be built for two thirds the cost of a tract house today.

An optimistic assessment, perhaps, but the development of earth shelter communities is galloping forward at breakneck speed. In Missouri real estate developer Ron Pfost has labelled earth shelter the biggest building boom he's seen since the apartment building spurt of the sixties. Determined to get in on the good thing, Pfost is putting together Kiva Ridge, a 110-acre, 35–40-unit subdivision. A successful apartment builder and manager, Pfost grew worried about the impact of spiralling energy costs on construction. He began to search for alternate methods that would survive the energy-parched days ahead.

At an energy conference held in Kansas City in 1979 Pfost discovered earth sheltering. An instant convert, he headed into the hills to find the perfect south-facing site. After a year of looking he found 110 acres near Kansas City. The slope and orientation were good. Pfost had his earth-sheltered promised land.

Michigan developer Dan Paulson followed the cautious road to earth-sheltered development. Before the first lines were on paper, he consulted with the bureaucrats who would rule on any development he would come up with. "These people see themselves as experts," he notes, "and want their expertise used. If you do your homework and know what you're talking about, then approach them as members of the same team without trying to impress or overwhelm them; their assistance can be highly valuable."

To get his development going, Paulson needed every bit of support and guile he could muster. He talked FHA into supporting the project and got subcontractors to join in advertising costs and his architects to defer payment until homes were sold. He found a sixteen-plot tract near Traverse City on which he plans to build custom-designed earth shelters, each home with a southern exposure. Prices will run from $60,000 to $100,000.

At Cerro Gordo, Oregon, one hundred families have banded together to create an environmentally sound community that will eventually house twenty-five hundred people. Plans include underground cluster homes or apartments, businesses, and schools. Private cars will be banned within the community. With energy requirements decreased by 50%, town planners figure they will be able to generate all the energy they need from sun, wind, water and biofuels. They even hope to generate their own income through small, nonpolluting town businesses. Members of the Cerro Gordo experiment include retirees, musicians, poets and a chemist.

Muir Woods, near Indianapolis, offers 132 condominiums, both above ground and below. The earth-sheltered units are bermed and earth-covered on the roof, with a garage buffering the northern wall. The heavily insulated homes are the first earth homes approved by Indiana's building commissioner.

In Minneapolis, the 12-unit Seward town houses show another use for earth shelter in urban settings. They are designed to make the best use of 110 × 330-foot urban lots. Energy Efficient Environments of Atlanta also has a 144-unit town house project on the way in Minneapolis, as well as developments in Arizona and California.

In New York, earth shelter is being used in a new student dormitory at the State University Agricultural and Technical College in Morrisville. Developments seem to have sprung up all over.

Potential projects in two other cities show the enormous potential of earth shelter for city development. It goes far beyond merely putting up country club estates down under. In Fort Worth, Texas, Frank Moreland generated a plan to incorporate earth design into the city land use plan. The city's housing is old and must be replaced if the community is to remain vital. While the detailed plan received local approval, the feds turned it down for funding. Still, the 1978 proposal that follows is most indicative of how earth shelter can help transform a city.

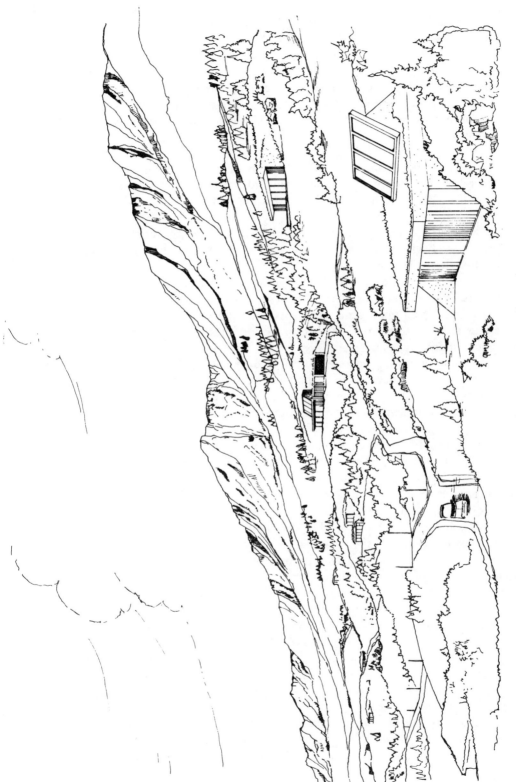

Malcolm Wells's vision of why earth shelter suburbs are better. The homes can be closer together than they are today and have far less impact on the land and the eye. The community remains a green, verdant park.

Four earth-sheltered prototypes will be constructed on sites selected by the city of Fort Worth. The sites will be in existing urban neighborhoods selected on the basis of their revitalization potential; each area will be supported by other federally funded programs and city commitment for the improvement of public facilities and services. The intent of this project is to revitalize such areas, and the process will include the improvement of the properties near the sites selected for the new earth-sheltered infill dwellings.

To support and encourage the investment by middle income buyers of the earth-sheltered houses, four nearby existing homes will be purchased and rehabilitated to middle income standards, in terms of size and amenities, as well as historical or cultural significance, and resold to middle income people.

The rehabilitation design will be consistent with energy conservation objectives and will be the responsibility of the same architect designing the earth-sheltered housing.

Experienced realtors will then market the homes to middle income buyers. The expectation is that the infusion of new housing, new green space and capital improvements in these neighborhoods will spark an interest in the private housing market to explore this alternative to traditional suburban development.

The goal of this proposed project is to encourage the revitalization of the inner city by developing marketable alternative middle income housing in lower middle income neighborhoods, while preventing or minimizing the relocation of present lower income residents.

In St. Paul, Minnesota, local officials are giving serious consideration to using earth shelter in conjunction with unfinished local highway sites. A number of abandoned highways lie around the Mississippi River city and earth shelters could put this wasted space to good use.

One abandoned highway, whose construction was blocked in 1972 by citizen protest, winds around a scenic bluff. In the years since, only the squirrels and dirt bikers have enjoyed the view, but recently St. Paul's city government asked for proposals to use the area. One proposal stated: "The Pleasant Avenues Corridor, with its bluff and grade embankements, offers an excellent site for construction of earth-sheltered housing. Much of the existing roadway grades can be incorporated into an earth-sheltered development.

"Development of earth-sheltered housing on the site," the report continues, "would create a major housing resource at considerable energy savings to the city." The proposal calls for some 1,770 units on 207 acres. A second St. Paul abandoned highway site sports 12 habitable acres. "We could have some of the most attractive underground housing by building into the bluffs," states the St. Paul city council president.

What will be the ultimate commercial impact of earth sheltering? Presumably it will be most significant. We are already somewhat behind other countries in using this new form of design. As urban economist Irving Hoch noted some years ago, "It is somewhat disquieting that in one hundred Russian cities, 35% or more of the investment in structures is underground. Do they know something we do not?"

In the final analysis, earth shelter can only have its maximum impact if it creates changes where today's building and design patterns are at their worst— and that's in the sprawl of our uncontrolled suburbs. It was developers who created this mess over the past couple of generations. And it will require earth shelter developers of similar clout and persistence but improved foresight to impose a new order on the concrete chaos.

PART FIVE

AN EARTH–SHELTERED FUTURE

If energy costs continue to go up, the United States must inevitably continue to go down under an earth blanket. The revitalized form of earth sheltering simply offers too many design and aesthetic advantages to be ignored in the years ahead.

The nation is littered with buildings constructed according to the attitudes and requirements of an utterly different period. In the past, materials were abundant and energy was cheap. Space was never a consideration—there was plenty of new land to gobble up out in the suburbs. Massiveness was considered a virtue and closeness with the surface was thought to be a quaint notion of the past.

Today, increasingly, these profligate, mass-oriented, energy-wasting predilections seem outmoded and just plain dumb. It no longer seems creative to build a huge steel-and-glass structure; it seems outrageous. Our times demand ideas that come from completely different directions than these.

Energy saving should be paramount in today's architectural styles. So should the avoidance of costly, scarce materials. Efficient land use counts for much, as do flexibility and style. In this inflationary age, designs should be inexpensive and long-lasting.

The form that encompasses all these attributes is earth shelter. It makes more sense for the present than any other style of housing. Virtually all the good ideas of solar design can be incorporated into it at greater energy savings than in an uncovered solar home. It uses cheap, abundant materials in a simple, sturdy-construction format. It can adapt to any environment and geography and can be designed to suit any style. It is an idea whose time has arrived.

Over the next few years earth shelter should become ever more popular. Our energy situation will probably grow steadily worse, making conventional homes less and less attractive. New energy-efficient surface homes will undoubtedly be tried, but earth shelter solves energy-efficiency problems far better than any surface system. It doesn't require a lot of hardware or add-on expense. The energy saving is built in and lasts as long as the building.

Some earth shelter experts predict that the energy crunch will cause older buildings to be retired and torn down sooner. Banks will become eager, rather than unwilling, to accept financing for sound energy-efficient designs, particularly earth shelters.

Steadily the form of our houses will change. The percentage of earth-sheltered buildings will rise each year. Within the decade they could possibly even come to predominate new home construction.

Rather than merely being desirable, earth sheltering may actually be necessary for the preservation of U.S. cities in the tight times ahead. "Our cities are the most inefficient, wasteful, hazardous and overdesigned in mankind's history," Frank Moreland has observed. "The United States has no future unless we rebuild our cities."

Cities that can offer its citizens surface-level amenities, such as parks, bike paths, open views, atriums, plazas and a mix of communal interaction and privacy are the most favored by Americans today. Earth sheltering can provide these much-desired surroundings at far less cost than any alternative. It can, and probably must, transform our urban vistas if we are to experience a better life.

Even back in the dark days when building underground brought to mind nothing so much as the thought of dank, desperate bomb shelters, the thought that shifting cities beneath the surface could be a cultural benchmark was a powerful one. In 1962, urban planner Milo D. Folley wrote: ". . . Although we see the subterranean environment as a protective shield in today's bomb shelter solutions, building below the ground may be a major step in our cultural development. From the evil of today comes the seed of tomorrow's culture."

Since that time the idea of earth shelter as a way to rid the world of its offensive, polluting urban sprawls has consistently grown stronger. Do skyscrapers really make sense in an age of scarcity? Do bland, ugly factories require space on the surface when they offer nothing appealing to the senses? What about the gross cubes of shopping centers? Wouldn't all this be better under a shielding layer of earth?

If technology and creativity have given us the tools to leave our surface much as nature intended it, without giving up the required functions of modern society, shouldn't we take advantage of the opportunity?

This is the true goal of many earth shelter designers. They are not just interested in the construction of a few odd houses or to gain the creation of an interesting but ineffectual model community. They want to use this new/old idea to re-create an earlier mode of American life. Seduced by the abundance of resources this continent provided, we have strayed from living in harmony with our surroundings. Earth shelter can lead us back to this symbiosis, where our greatest strengths since our formation as a nation can be found. Living in concert with the earth is where earth shelter designers feel we belong. The land itself forms the grandest architecture possible. Everything humans do merely cheapens the beauty and limits the ability of nature. As Arthur Drexler has observed, the land-contour models crafted by architecture students are often the most lovely things they make, far more arresting than any structure they will later lard onto the site. These land studies, he argues, "are extremely beautiful.

A fanciful Malcolm Wells notion for an earth-sheltered office park. Earth would cover the roof of each floor, thus saving energy, and an underground mall would extend out to the ground-level skylights.

Why shouldn't they be? Often the land is very beautiful. These contour maps
—these models of the earth—are already architecture. Nothing else is needed to
make a building, you know, except to pull out one or two of these layers and
make a space between them. Think how many thousands of buildings in the
United States could slip into the earth. Instead, the students think of architecture as the making of things in opposition to the earth."

The leading visionary of an earth-sheltered future is Malcolm Wells. He
longs for the creation of an earth national park, where the surface is reserved
for recreation and pleasure, with the business going on below. He foresees
green cities developing slowly, piece by piece. These earth-sheltered urban
zones will still have high-rises, but the miles and miles of drab low-rises and factories will be gone. Wells can't see any reason why we can't have efficient housing that will turn the earth into a park again.

It's no dream, either. After all, we're running out of land, and earth shelter
represents far better land use than what we have been doing. It provides for the
possibility of high-density housing plus privacy for an increasingly crowded
population—without destroying the beauty of the surface.

Imagine New York City composed almost entirely of parks and playgrounds,
open spaces and promenades, instead of the seemingly endless rows of bland
brick boxes that make it up today. Through earth shelters that barely extend
below surface level, all the people living in the borough of Queens could be fit
into a fraction of the available space. They could live closer together than they
now do, with more privacy and less noise. The freed land could be used to
bring the residents joy, rather than being used for barracks. It would not turn
the city into the country; it would make a better brand of city.

Beyond this, Wells envisions earth shelter permitting a flexibility of design and
use impossible to achieve with present methods. Why not manufacture simple
concrete components that can be plugged into and removed from permanent
earth-sheltered shells? If you need a hospital during an emergency, simply cart
in hospital units and plug them in. If you have a growing student population,
plug in extra school units or living units or whatever you require. We'll be able
to simply unplug the school components, send them to the recycling center and
slide the hospital parts—completely furnished—into the shell. In the not-too-
distant future the plug-in parts may even be solar-powered and so well-
insulated they'll lose almost no heat in wintertime. They could transform our
organic wastes into soil-enriching nutrients and reclaim all their water. With
even better luck we may someday learn to grow great earth shells the way we
now grow crystals in a test tube. Imagine huge, shimmering tetrahedrons growing one- and two-hundred-foot clear spans in the sides of hills! Now *that's* organic architecture!

The melding of earth-sheltered ideas with a reverence for nature can be incorporated in a virtual philosophy of life based on the new way of thinking about
dwellings. The form of our buildings should not derive from what we want of
them but from what nature offers us as an ideal. We should then use the form
Nature provides, bending ourselves to her rather than the other way around.

We should not go under the soil for novelty or protection or energy savings.

Some of these are good reasons in and of themselves, but they are not good enough. We should work with earth shelter because it puts us in contact with—in harmony with—our natural world. It will foster in us a new awareness of, and respect for, that world. We are dependent upon the cycles and order of the natural world and yet everything we do today fights that order or tries to deny it. Only when we accept and work with nature will we achieve forms of the greatest possible grace, permanence and sanity. We can achieve this through earth shelter if we develop the new form with a proper attitude.

In the words of Malcolm Wells, "It will all be seen as one many-faceted issue: reverence for life. That's when we'll start to see some stunning buildings. Architecture is one expression of human values and attitudes. When they get straightened out, you can expect to see houses—even whole cities—as beautiful-and appropriate-looking as the loveliest of wild landscapes. But we're not there yet, not by a long shot."

Even earth-sheltered high-rises aren't out of the question. Designers could combine surface ambience with high-rise density.

An earth-sheltered future might see the creation of movable concrete living units that could be plugged into and out of earth-covered service centers.

Skiopolis, a concept of Washington architect Roy Mason, shows how earth shelter ideas can be used to create technological environments that do not adversely affect the appearance of the surface. Earth shelters don't need to look like traditional buildings at all—and in time they probably won't. The interior of Skiopolis would hold a residential development and shopping facilities.

APPENDIX ONE:

EARTH SHELTER ORGANIZATIONS

The following groups offer the best source of information on recent developments in the field and on locating experts in your geographical area.

Clearinghouse for Earth Sheltered Building
School of Architecture and Environmental Design
University of Texas at Arlington
Arlington, TX 76019

Earth Shelter Builders Association
P. O. Box 19148
Minneapolis, MN 55419

The Underground Space Center
11 Mines and Metallurgy Building
221 Church St. S.E.
Minneapolis, MN 55455

APPENDIX TWO:

EARTH SHELTER PERIODICALS

Earth Shelter Digest & Energy Report
WEBCO Publishing, Inc.
1701 E. Cope St.
St. Paul, MN 55109
Wide-ranging, informative and entertaining review of the earth-sheltered scene; particularly strong on case studies. Bimonthly. Subscription: $15 a year; $25 for two years.

Underground Space
Journal of the American Underground Space Association
Dept. of Civil and Mineral Engineering
University of Minnesota
Minneapolis, MN 55455
Sometimes dry and technical, but an unmatched reservoir of expert opinion. Bimonthly. Subscription: $62 a year; $117.80 for two years.

The following magazines also deal with earth shelter topics fairly regularly:

American Institute of Architects Journal
1735 New York Avenue N.W.
Washington, DC 20006

Architectural Record
McGraw-Hill Publications
Rockefeller Center
New York, NY 10020

The Mother Earth News
105 Stoney Mountain Rd.
Hendersonville, NC 28739

Popular Science
P. O. Box 2871
Boulder, CO 80302 (circulation office)

Progressive Architecture
600 Summer St.
Stamford, CT 06904

Solar Age
Church Hill
Harrisville, NH 03450

APPENDIX THREE:

ARTICLES ON EARTH SHELTERING

GENERAL

Confessions of a Gentle Architect, Malcolm Wells, *Environmental Quality*, July 1973.

Conservation Architecture, Malcolm Wells, *Architectural and Engineering News*, September 1969.

Conservation of Energy by Use of Underground Space, Thomas P. Bligh and Richard Hamburger, National Academy of Sciences, 1974.

Down to Earth, Susan Slesin, New York *Times*, March 22, 1979.

Down to Earth Housing, Wybe J. Van der Meer, *Solar Age*, September 1976.

Down Under, Down Under . . . Or How Not to Build Underground, Malcolm Wells, *Progressive Architecture*, March 1968.

The Earth's the Ceiling, James Gorman, *The Sciences*, March/April 1976.

The Economics of Building Down, Michael de Courcy Hinds, New York *Times*, March 22, 1979.

Energy Conservation by Building Underground, Thomas P. Bligh, *Underground Space*, May/June 1976.

Energy Conservation Forces a Revolution in Buildings, *Engineering News-Record*, November 3, 1975.

Energy Essays, Malcolm Wells, Edmund Scientific Company, 1976.

Geotecture, Royce LaNier, Library of Congress Catalog Card Number 76–139958, 1970.

Geotecture: Concept, Design, Construction and Economy of Geospace—The Creation of Subterranean Accommodation, Patrick Hosburgh, University of Notre Dame, 1973.

Going Under to Stay on Top, Charles Fairhurst, *Underground Space*, July 1976.

The Good Feeling of Living in the Earth, Ray Wolf, *Organic Gardening*, December 1978.

The Good Life Underground, Mike Edelhart, *OMNI*, January 1980.

Houses: The Architect Speaks to Man's Needs, *Progressive Architecture*, May 1974.

Is It Time to Go Underground?, Lt. Lloyd Harrison, Jr., *The Navy Civil Engineer*, Fall 1973.

Living Underground, *Newsweek*, June 5, 1978.

New Cave Dwellers Save on Energy Bills, Keep Comfortable, Shirley A. Jackewicz, *Wall Street Journal*, May 24, 1979.

New Homes Revive the Ancient Art of Living Underground, David Martindale, *Smithsonian*, February 1979.

Non-Traditional Military Uses of Underground Space, Lloyd S. Jones, *Underground Space*, November 1977.

Nowhere to Go but Down, Malcolm Wells, *Progressive Architecture*, February 1965.

Saving by Going Underground, *American Institute of Architects Journal*, February 1974.

Underground Construction, *Buildings*, January 1977.

Underground Housing, R. F. Dempewolff, *Science Digest*, November 1975.

Underground Living, V. Elaine Smay, *Popular Science*, June 1974.

What Your Home Will Be Like in Fifty Years, Malcolm Wells, *Mainliner*, September 1974.

Why I Went Underground, Malcolm Wells, *The Futurist*, February 1976.

LAND USE

Earth Covered Habitat—An Alternative Future, Frank L. Moreland, *Underground Space*, August 1977.

Earth Shelter Architecture and Land Use Policy, Royce LaNier and Frank L. Moreland, *Underground Space*, August 1977.

Liberating Land: A Blueprint for Urban Growth, Gunnar Birkerts, *Progressive Architecture*, March 1973.

Planning the Underground Uses, Donald F. Hagman, National Academy of Sciences, 1974.

Subterranean Urban Systems, Gunnar Birkerts, University of Michigan, 1974.

DESIGN

The Architectural Underground, Part I, Kenneth Labs, *Underground Space*, May/June 1976.

The Architectural Underground, Part II, Kenneth Labs, *Underground Space*, July/August 1976.

The Energy Factor—A Dimension of Design, David Bennett and Thomas P. Bligh, *Underground Space*, August 1977.

Subsurface and Underground Housing in Hot, Arid Lands, U. S. Army Research Office, 1964.

Underground Architecture, Malcolm Wells, *CoEvolution Quarterly*, Fall 1976.

Underground—The Prairie Sod House Returns, *Architecture Minnesota*, September/October 1977.

CASE STUDIES

An Instinctive Home, *Progressive Architecture*, May 1964.

The Beale Solar-Heated Subterranean Guest House, *Mother Earth News*, May/June 1977.

Digging In: How a Family Lives Underground, T. L. Gettings, *Organic Gardening*, June 1978.

Go Underground Young Man, Says John Strickler, Linda Lewis, Seattle *Post-Intelligencer*, March 6, 1977.

Going Underground: House of Tomorrow, Bonnie Speer, *The Sunday Oklahoman*, April 10, 1977.

Living History Farms, the Farmhouse of Today and Tomorrow, Ray D. Crites, Ames Design Collaborative, September 1976.

My Cave, Andy Davis, *Underground Space*, November 1977.

Plowboy Interview, Andy Davis' Cave, *Mother Earth News*, July/August 1977.

The Self-Heating, Self-Cooling House, Wendell Thomas, *Mother Earth News*, Reprint Number 39.

Solaria, Bob and Nancy Homan, Dr. Harry Thomason, Malcolm Wells, Edmund Scientific Company, April 1974.

Underground Houses, V. Elaine Smay, *Popular Science*, August 1977.

Underground Living in this Ecology House Saves Energy, Cuts Building Cost, Preserves the Environment, V. Elaine Smay, *Popular Science*, June 1974.

Why We Moved Underground, Frank Booker, *Farmstead*, March 1979.

Winston House, Lyme, New Hampshire, Designer and Contractor Don Metz, *Progressive Architecture*, May 1975.

Your Next House Could Have a Grass Roof, R. F. Dempewolff, *Popular Mechanics*, March 1977.

EARTH SHELTER AND SOLAR

How to Trap Solar Heat with Your Windows, Edward Allen, *Popular Science*, February 1975.

Lee Porter Butler: He Builds Houses that Never Cool Off, *Hudson Home Guide: Home Building and Remodeling*. Oradell, New Jersey, November 1979.

Living Underground (four articles), *Solar Age*, October 1979.

Passive Solar Goes Underground, V. Elaine Smay, *Popular Science*, April 1978.

Solar Information Directory, *Solar Age*, August 1980.

PUBLIC POLICY

Insuring Risks Underground—Some General Considerations, E. M. De-Saventhem, *Underground Space*, September 1977.

Legal Aspects of the Use of the Underground, Dan A. Tarlock, *National Academy of Sciences*, 1974.

APPENDIX FOUR:

BOOKS FOR FURTHER READING

GENERAL

Alternatives in Energy Conservation: The Use of Earth Covered Buildings. Classic collection of papers from the 1975 Fort Worth earth shelter conference. The National Science Foundation backed publication of the proceedings. Order from the U. S. Government Printing Office, Washington, DC 20402. Stock Number 038-000-002864. $3.25. Library of Congress classification 76–22819.

Alternatives in Habitat: The Use of Earth Covered Settlements. Frank Moreland, editor. Proceedings of the May 1978 World Earth Shelter Conference. Order from: National Technical Information Service, U. S. Department of Commerce, 5285 Port Royal Rd., Springfield, VA 22116.

Architectural Use of Underground Space, Ken Labs. A masterful master's thesis that ties together the basic concepts of the movement perhaps better than any other work. Order from: *Earth Shelter Digest.* $21.

Earth Sheltered Housing. Transcript of a workshop held at the Montana Department of Natural Resources and Conservation in July 1978. Covers a broad range of topics. Request from: Department of Natural Resources and Conservation, 32 S. Ewing St., Helena, MT 59601.

Earth Sheltered Living—An Introduction. Economic analysis of the savings possible through earth shelter. Order from: *Earth Shelter Digest.* $6.50 plus $1 handling.

Gentle Architecture, Malcolm Wells. An overview of the field by the conscience of earth sheltering. McGraw-Hill. $22.50.

Housing Alternatives. A booklet from Cave Enterprises summarizing the positive points of earth shelter in an energy-starved age. Order from: *Earth Shelter Digest*, 1701 E. Cope St., St. Paul, MN 55109. $3.

The Underground House Book, Stu Campbell. Garden Way Publishers, Vermont. Folksy, rural-oriented look at earth shelter houses. $9.95.

Underground Utilization: A Reference Manual of Selected Works, Truman Stauffer. Eight volumes covering every aspect of earth sheltering. For information write: Department of Geosciences, University of Missouri–Kansas City, Kansas City, MO 64110.

LAND USE

Potential Use of Underground Space, Department of Civil and Mineral Engineering, Minneapolis Campus, University of Minnesota, Feb. 1975, 75 pages. $7.

Preliminary Design Information for Underground Space, Department of Civil and Mineral Engineering, Minneapolis Campus, University of Minnesota, August 1975, 98 pages. $7.

DESIGN

Designs for Small Atrium Homes, Earth Habitats, P. O. Box 401072, Dallas, TX 75240.

Earth Homes, Richard Ohanian. A jam-packed display of the home styles used by Shelterra, an Ohio earth shelter builder. It's $14.95 from: *Earth Shelter Digest*.

Earth Integrated Architecture, edited by and available from: James W. Scalise, College of Architecture, Arizona State University, Tempe, AZ, 1975. $10.

Earth Integrated Building Construction, Portland Cement Association. A major earth shelter supplier goes over structural problems. Order from: Portland Cement Association, 5420 Old Orchard Rd., Skokie, IL 60077. Free.

An Earth Sheltered Guest House, Oklahoma State University. Designs by students for an earth-sheltered guest house, some with unusual approaches. Available for $8 from: Architectural Extension, 115 Architecture Building, Oklahoma State University, Stillwater, OK 74074.

Earth Sheltered Housing Design, The American Underground Space Association, Van Nostrand Reinhold. The definitive review of the structural questions and design alternatives for earth sheltering. Order from: Van Nostrand Reinhold, New York, NY. $10. Malcolm Wells sells the book for $12 together with his critique of certain portions he disagrees with.

Homes In the Earth, Jeremy A. Jones. Ideas from Design Concept Associates in Spokane, Washington. Available for $7.95 from: *Earth Shelter Digest.*

Subterranean Homes, Joe Hylton. A collection of floor plans and design ideas from an experienced earth shelter architect. This privately published book can be ordered for $12, plus $1.50 handling, from: *Earth Shelter Digest.*

Sun Belt Earth Sheltered Architecture, John Langley and James Gay. A regional review. Order from: P. O. Drawer 729, Winter Park, Florida 32790. $7.

Terracture, J. W. Scalise, published by Arizona State University, 1974.

Underground Designs, Malcolm Wells, 1977, 87 pages, $6. Available from: Malcolm Wells, P. O. Box 1149, Brewster, MA 02631. Plans and illustrations of Wells's designs, both practical and fanciful, along with sections on selecting a site, choosing a structural form, coping with code problems, waterproofing, insulation and landscaping.

The Underground Plans Book, Malcolm Wells and Sam Glenn-Wells. $13. A giant-size collection of plans and hints from the earth shelter pioneer. The designs aren't meant for building but for stimulating ideas that can be translated into personal designs. Excellent. Order from: Wells, P. O. Box 1149, Brewster, MA 02631.

CASE STUDIES

The Alternative House, Rita Tatum, Reed Books, 9155 W. Sunset Blvd., Los Angeles, CA 90069. $6.95. Contains a solid chapter on earth shelters, especially those of William Morgan.

A Case for Underground Schools, Oklahoma State University, Department of Education, Oklahoma City, OK.

The $50 and Up Underground House Book, Mike Oehler, Mole Publishing, $5. A do-it-yourselfer with a nice, light writing touch tells how he did it. Low-tech in the extreme, but might give you an idea or two. Order from: Mole Publishing, Route 1, Box 618, Bonners Ferry, ID 83805.

Go Underground and Save, Gary Rickman. Author tells you how he built an earth shelter near Wellsville, Kansas. Striking design concept. $4.95 from Underground Homes, RR 1, Box 160D, Wellsville, KS 66092.

Underground Houses: How to Build a Low Cost Home, Rob Roy, Sterling Publishers, $5.95. A New York back-to-the-lander goes over the unique plans he followed in building a log earth shelter. Order from: Sterling Publishers, 2 Park Ave., New York, NY 10016.

CONSTRUCTION

From the Ground Up, John N. Cole and Charles Wing, Atlantic Monthly Press, Little Brown and Co., 1976, 244 pages. Information and illustrations intended to provide guidance on designing and building an energy-efficient shelter. Not specifically on earth-sheltered houses.

The Owner-Builder and the Code, Ken Kern, Ted Kogon and Rob Thallon. Not specifically about earth shelter, but first-rate information about coping with codes. Good even when working with an architect, since it will help you deal with banks from position of knowledge. $5 from: Owner-Builder Publications, P. O. Box 550, Oakhurst, CA 93644.

The Owner Built Home, Ken Kern, Scribners. Contains detailed suggestions on how to build an inexpensive functional home yourself. Discusses areas such as site and climate, materials and skills, form and function, and design and structure.

Means' Building Construction Cost Data, Means Co., $24.50. A yearly review of construction costs in minute detail. Use it to figure the expense for various parts of your house. Write to: R. S. Means Co., 150 Construction Plaza, Kingston, MA 02364.

EARTH SHELTER AND SOLAR

All Through the House. A Guide to Home Weatherization, Thomas Blandy and Denis Lamoureux, 184 pages, illustrated. McGraw-Hill. $7.95 (paper).

At Home in the Sun, Norah Deakin Davis and Linda Lindsey, 236 pages, illustrated. Garden Way Publishing, Charlotte, VT 05445. $9.95 (paper).

Autonomous House Report, Nick Nicholson. Three volumes, illustrated, paper. Part One: 48 pages, $4.95; Part Two: 64 pages, $5.95; Part Three: $5.95; the set, $16, including postage. Order from: The Ayer's Cliff Centre, Box 344, Ayer's Cliff, Quebec, Canada JOB 1CO.

Country Energy Plan Guidebook, Alan Okagaki and Jim Benson. Institute for Ecological Policies, Fairfax, VA. $7.50 for individuals and public-interest groups; $15 for all others.

Designing and Building a Solar House: Your Place in the Sun, Donald Watson, Garden Way Publishing, Charlotte, VT 05445, 1977, 284 pages. $8.95.

Energy-Efficient Community Planning. A Guide to Saving Energy and Producing Power at the Local Level. James Ridgeway, 221 pages, illustrated. JG Press, Emmaus, PA. $14.95 (cloth), $9.95 (paper).

From the Walls In, Charles Wing, 226 pages, illustrated. Atlantic Monthly Press Book. $9.95.

A Golden Thread: 2500 Years of Solar Architecture and Technology, Ken Butti and John Perlin, 320 pages, illustrated. Cheshire Books, 514 Bryant St., Palo Alto, CA 94301. $15.95 (cloth). Co-published with Van Nostrand Reinhold.

How to Use Solar Energy in Your Home and Business, Ted Lucas, Ward Ritchie Press, Pasadena, CA, 1977, 315 pages. $7.95. This book contains detailed information on different solar heating systems to be used for different heating and cooling needs. It also discusses passive solar heat and conservation techniques.

An Inexpensive, Economical Solar Heating System for Homes, J. W. Alfred, J. M. Shinn, Jr., E. Kirby, and S. R. Barringer, Langley Research Center, Hampton, VA 23665, July 1976, 56 pages, NASA TM-X-3294. This report describes a low-cost solar home-heating system to supplement present warm-air heating systems.

Natural Solar Architecture: A Passive Primer, David Wright, Van Nostrand Reinhold, $7.95. A lusciously illustrated volume that imparts a deep sense of environmental understanding of the relationship between house, sun and earth. Order from: Van Nostrand Reinhold, 135 West 50th St., New York, NY.

Passive Solar Award Homes, Department of Housing and Urban Development. Free. A collection of 162 passive solar designs that the government feels are worthy of recognition. Some include earth-sheltered features, but all might give you ideas you could adapt to in-ground construction. To receive a free copy contact HUD, P. O. Box 280, Germantown, MD 20767; 301-251-5154.

Passive Solar Energy, Bruce Anderson and Malcolm Wells, Brick House Publishing, 3 Main St., Andover, ME 01810.

Solar Access Law, Gail Boyer Hayes, 303 pages. Ballinger Publishing Co., Cambridge, MA 02138. $18.50 (cloth).

The Solar Cat Book, Jim Augusty. Illustrated by Hildy Paige Burns. 96 pages. Ten Speed Press, P. O. Box 7123, Berkeley, CA 94707. $3.95 (paper).

Solar Dwelling Design Concepts, U. S. Department of Housing and Urban Development by the AIA Research Corporation, Washington, DC. Contract IAA H-5574, 146 pages, May 1967.

Solar Gain: Winners of the Passive Solar Design Competition, California Energy Commission, Publications Unit. Illustrated, 119 pages. $3.95 (paper), includes postage. Suite 616, 1111 Howe Ave., Sacramento, CA 95825.

The Solar Home Book—Heating, Cooling and Designing with the Sun, Bruce Anderson with Michael Riordan, Cheshire Books, Harrisville, NH, 1976, 297 pages. $8.50.

Thermal Delight in Architecture, Lisa Heschong, 78 pages. The MIT Press, Cambridge, MA 02142. $5.95 (paper).

30 Energy Efficient Houses . . . You Can Build, Alex Wade and Neal Ewenstein, Rodale Press, Emmaus, PA, 1977, 315 pages. $8.95.

PUBLIC POLICY

Legal, Economic and Energy Considerations in the Use of Underground Space, National Academy of Sciences, Washington, DC, 1974. Papers from the June 1973 Engineering Foundation Conference, Berwick Academy, South Berwick, ME.

Need for National Policy for the Use of Underground Space, Engineering Foundation Conference, Berwick Academy, South Berwick, ME, June 1973, 232 pages. A collection of conference papers.

Earth Sheltered Housing: Code, Zoning and Financing Impediments, U. S. Department of Housing and Urban Development, Washington, DC. Free. A quick review of the problems.

BIBLIOGRAPHY

Energy Inform. For $5 these folks will send you a complete bibliography of earth shelter writings. Energy Inform, 3528 Dodge, Omaha, NE 68131.

HISTORY

Prehistoric Architecture in the Eastern United States, William Morgan, MIT Press. $25. One of the most innovative earth shelter architects traces the roots of subsurface living in early America. Fascinating stuff. Order from: MIT Press, Cambridge, MA 02142.

APPENDIX FIVE:

EARTH SHELTER MOVIES

Grass On the Roof. A half-hour look at earth-sheltered houses. 16mm color. Can be bought or rented. For information: Filmart Productions, Inc., 199 E. Annapolis St., St. Paul, MN 55118.

Terraset's Sun. This film follows the birth of an earth-sheltered school in Reston, Virginia. The 16mm color movie is twenty minutes long. It can be purchased for $345 from: Webco Publishing, Inc., 479 Fort Rd., St. Paul, MN 55102.

APPENDIX SIX:

EARTH SHELTER DESIGNERS

No list of earth shelter designers can be totally complete, since new people enter the field each day. Earth shelter architecture, remember, is a style, not a discipline. Any registered architect with enough enthusiasm can design an earth-sheltered structure. Those listed here concentrate heavily or totally on this type of building.

ARCHITECTS

John Barnard
1054 Main St.
Osterville, MA 02655

Lester Boyer
Professor of Architecture
103 Architecture Bldg.
Oklahoma State University
Stillwater, OK 74074

William Chaleff
P. O. Drawer XXXX
East Hampton, NY 11937

Roland Coate
1905 Lincoln Blvd.
Venice, CA 90241

David Deppen
P. O. Box 3945
Portland, OR 97208

Joe Hylton
566 Buchanan Ave.
Norman, OK 73069

Philip Johnson
375 Park Ave.
New York, NY 10022

Kenneth Labs
Undercurrent Design Research
147 Livingston St.
New Haven, CT 06511

John Langley
P. O. Box 729
Winter Park, FL 32790

Vince LaTona
11301 Manchester St.
Kansas City, MO 64134

Roy Mason
2011 R St. NW
Washington, DC 20009

Ron McClure
Box 628
Tigeras, NM 87059

Michael McGuire
432 S. Main St.
Stillwater, MN 55082

Don Metz
Box 525
Lyme, NH 03768

Frank Moreland
Center for Energy Policy Studies
University of Texas at Arlington
Arlington, TX 76010

William Morgan
220 E. Forsythe St.
Jacksonville, FL 32202

James Scalise
College of Architecture
Arizona State University
Tempe, AZ 85281

David Wright
Sea Group
P. O. Box 49
Sea Ranch, CA 95497

Malcolm Wells
P. O. Box 1149
Brewster, MA 02631

DESIGN FIRMS

Architectural Alliance
400 Clifton Ave. South
Minneapolis, MN 55403

Berg & Associates
1315 Garland Lane
Wayzata, MN 55391

Big Outdoor People, Inc.
26600 Fallbrook Ave.
Wyoming, MN 55092

Carmody & Ellison
1800 Englewood Ave.
St. Paul, MN 55104

Coffee & Crier
509 Oakdale Dr.
Austin, TX 78745

Colorado SunWorks
959 Walnut St.
Boulder, CO 80302

Concept 2000, Inc.
19003 N. 52nd Ave.
Glendale, AZ 85308

Design Concept Associates
North 14 Howard, Suite 303
Spokane, WA 99201

Earth Habitats
P. O. Box 401072
Dallas, TX 75240

Earth Shelter Designs
P. O. Box 233
Marine on St. Croix, MN 55047

Earth Sheltered Home Designers
P. O. Box 1191
Girardeau, MO 63701

ECOC Corporation
New Britain Ave.
P. O. Box 331
Farmington, CT 06032

Energy Efficient Environments
173 W. Wyenca St., Suite 205
Atlanta, GA 30342

Everstrong, Inc.
Route 3, Box 55-B
Redwood Falls, MN 56283

Geobuilding Systems/Jay Swayze
P. O. Box 1556
Hereford, TX 79045

Hanscomb Associates
600 W. Peachtree St. NW
1038 Life of Georgia Tower
Atlanta, GA 30308

Shelterra Earth Homes
P. O. Box 967-A
Springfield, OH 45501

Solar Earth Consultants
2294 Weldon Parkway
St. Louis, MO 63141

Solar Earth Energy/Don Secrist
2020 Brice Rd.
Reynoldsburg, OH 43068

R. D. Strayer, Inc.
5490 Ashford Rd.
Dublin, OH 43017

Terra-Dome Underground Homes
14 Oak Hill Cluster
Independence, MO 64057

Underground Homes
700 Masonic Building
Portsmouth, OH 45662

U. S. Systems
496 Railroad Ave.
P. O. Box 955
Logan, OH 43138

Earth Shelter Corporation of
America, Inc.
Rt. 2, Box 97B
Berlin, WI 54923

Earth Systems
P. O. Box 35338
Phoenix, AZ 85069

Everstrong Marketing, Inc.
Box 351
Morton, MN 56270

Simmons & Sun, Inc., and Solar Earth
Consultants, Inc.
P. O. Box 1497
High Ridge, MO 73049

Tecton Corporation
111 W. Fillmore
Colorado Springs, CO 80907

Terra-Dome Corporation
14 Oak Hill Cluster
Independence, MO 64057

U'Bahn Earth Homes
4008 Braden St.
Granite City, IL 62040

DESIGN FIRMS THAT SELL FRANCHISES

American Solartron
Rt. 5, Box 170
Centralia, IL 62801

R. A. Burnett and Associates, Inc.
Rt. 2, Box 226
Rogersville, MO 65742

Davis Caves, Inc.
200 W. Monroe St.
Chicago, IL 60606

The concentration of well-known earth shelter organizations, as you can see, is in Oklahoma, Minnesota, Missouri and New England. To find an architect interested in going under the soil closer to home, you can try these information sources:

American Institute of Architects
1735 New York Ave. NW
Washington, DC 20006

Earth Shelter Data Bank
Earth Shelter Digest & Energy
   Report
1701 E. Cope St.
St. Paul, MN 55109

Conservation and Renewable
   Energy Inquiry and Referral
   Service
1-800-523-2929
1-800-462-4983 (Pennsylvania)
1-800-523-4700 (Alaska and Hawaii)

Also check with local libraries, architectural groups, architecture schools and the various media to see if they know of anyone who is heading into the earth.

APPENDIX SEVEN:

SOLAR INFORMATION

This U. S. Department of Energy list covers community groups concerned with solar utilization and other energy possibilities. They might prove to be good sources of information for locating an earth shelter builder.

ALABAMA

James Horton
THA CETA Solar Project
4805 ⅜B Second St., East
Tuscaloosa, AL 35401
(205) 553-2485
Advice on organizing new community projects; information on solar greenhouses and window boxes.

Carlyle Pollock
Solar Greenhouse Employment
   Project
P. O. Box 1916
University, AL 35486
Workshops on solar greenhouses for small farmers and low-income organizations.

ALASKA

Alaska Federation for Community
   Self-Reliance
P. O. Box 73488
Fairbanks, AK 99707
(907) 456-7674
Energy audits; alternative energy information; referral services; community gardens.

Chris Noah
Alaska Council on Science and
   Technology
Pouch Ave.
Juneau, AK 99811
(907) 465-3510
Grants for appropriate-technology projects; Northern Technology Grants Program.

ARIZONA

Jim Gonzales
Arizona Community Action Assoc.
1001 North Central #312
Phoenix, AZ 85004
(602) 252–6067
*Training and technical assistance in
neighborhood-based technologies.*

ARKANSAS

Jim Deal
Arkansas Solar Coalition
1145 West Hearn St.
Blytheville, AR 72315
(501) 762–2769
*Networking and help in organizing
projects; demonstrations; workshops;
exhibitions for do-it-yourselfers.*

Beverly Maddox
National Weatherization Training
   Project
Div. of Community Services
1 Capitol Mall
Little Rock, AR 72201
(501) 370–5208
*Training project for all state
weatherization grantees across the
country.*

Edd Jeffords
Ozark Institute
Box 459
Eureka Springs, AR 72632
(501) 253–7384
(800) 632–0097 (in Arkansas)
*Training and technical assistance for
community action agencies; rural
producer co-ops in conservation and
appropriate technology.*

CALIFORNIA

Tom Tomasi
Office of Appropriate Technology
1530 10th St.
Sacramento, CA 95814
(916) 445–1803

*Assistance to local governments and
community organizations for energy
and economic development projects.*

Judy Corbett
SolarCal
1111 Howe Ave.
Sacramento, CA 95825
(916) 920–7621
*Technical assistance in local energy
planning and implementation;
networking.*

California Western SUN
6022 West Pico Blvd. #1
Los Angeles, CA 90035
(213) 852–5135
*Small grants; referrals for technical
assistance; community project
development assistance; job
development assistance.*

COLORADO

Bill Schroer
Colorado Energy Advocate Office
1020 15th St., #300
Denver, CO 80202
(303) 832–3291
*Referrals for low-income
organizations to sources of technical
assistance.*

CONNECTICUT

Bruce Wilbur
Co-op Extension Service
University of Connecticut
U36
Storrs, CT 06268
(203) 486–4129
*Energy outreach; workshops;
literature.*

Betsey Hedden
League of Women Voters of
   Connecticut
60 Connolly Parkway
Hamden, CT 06514
(203) 288–7996

Speakers; limited amounts of research materials.

Roger Gregoire
Center for the Environment and Man
  (CEM)
275 Windsor St.
Hartford, CT 06120
(203) 549–4000, ext. 309
Technical-assistance consultants to low-income organizations in solar and conservation fields.

DELAWARE

Carol Crouse, Chairperson
Delmarva Solar Committee
Box 691
Blades, DE 19973
(302) 629–6421
Technical information and networking for solar field.

H. Earl Roberts
Delaware Technical Community
  College
Terry Campus
1832 North Dupont Parkway
Dover, DE 19901
(302) 736–5401
Home energy conservation programs in Kent County; contacts at other Delaware Technical campuses.

DISTRICT OF COLUMBIA

Vince Kosker
Anacosta Energy Alliance
2027 Martin Luther King Dr. S.E.
Washington, DC 20020
(202) 899–7932
Information on weatherization, energy audits, solar installations; workshops.

FLORIDA

Dawn Tucci
L.E.A.P.
Governor's Energy Office
301 Bryant Bldg.

Tallahassee, FL 32301
(904) 488–6764
Community-based energy conservation program.

Greg Glass
Florida Assoc. of Community Action
  Agencies
P. O. Box 1775
Tallahassee, FL 32302
(904) 222–2043
Conducts a series of technical-assistance workshops throughout the state; appropriate technology.

Joan Partington
Florida Solar Coalition
935 Orange Ave.
Winter Park, FL 32789
(305) 647–0467
Citizens' grassroots organization that provides referral services and works with community-based appropriate technology.

GEORGIA

Betty Terry
Georgia Solar Coalition
3110 Maple Dr., Room 403 A
Atlanta, GA 30305
(404) 231–9994
Speakers; lending library; newsletter; workshops and seminars.

Ralph Paige
Federation of Southern Co-ops
40 Marietta, N.E. #1710
Atlanta, GA 30303
(404) 524–6882
Works with co-ops to provide information on appropriate technology, wood heat, and solar greenhouses.

HAWAII

Hawaii Natural Energy Institute
2540 Dole St.
Honolulu, HI 96822
(808) 948–8890

Speakers; newsletter; referral services; research and development in renewable and alternative energy sources.

Jim Harpstrite
Energy Education Project
University of Hawaii
1776 University Avenue
Honolulu, HI 96822
(808) 948–6831

IDAHO

Dan Smith
Idaho Solar Energy Assoc.
Box 2761
Boise, ID 83701
(208) 336–1526
Workshops in appropriate technology and conservation; networking.

Jim Worstell
Energy Specialist
Co-op Extension Service
University of Idaho
P. O. Box 300
Boise, ID 83701
(208) 334–2142
Workshops and seminars; material on alcohol-based fuels and other alternative technologies for farm communities.

ILLINOIS

Jeff Mitchell
Dept. of Commerce and Community
   Affairs
325 West Adams (4th fl.)
Springfield, IL 62706
(217) 785–2264
Technical assistance to local governments, small businesses and communities; aid in locating funding.

Art Rasch
Co-op Information Center
P. O. Box 2559
Chicago, IL 60690

(312) 227–5897
Assistance in establishing community-run co-ops in energy, food, etc.; some books and pamphlets; lists of suppliers.

INDIANA

Jean Warren
Alternative Technology Assoc.
P. O. Box 27246
Indianapolis, IN 46227
(317) 784–7744
Information on community-based technologies.

Lawrence Mayberry
Citizen Action Coalition
3620 North Meridian St.
Indianapolis, IN 46208
(317) 923–2494
Networking; assistance in organizing energy projects.

IOWA

Linda Nicholson
Energy Research and Information
   Foundation
3500 Kingman Blvd.
Des Moines, IA 50311
(515) 277–0253
Renewable energy workshops through community colleges; publications; library facilities.

Mike Coyne
Iowa Development Commission
250 Jewett Bldg.
Des Moines, IA 50309
(515) 281–3459
Technical assistance in community production projects (particularly alcohol-based fuels).

Skip Laitner
Community Action Research Group
   (CARG)
P. O. Box 1232
Ames, IA 50010

(515) 292–4758
*Information for local government and
community groups to organize and
implement projects.*

KENTUCKY

Don Huesman
Appalachian Science in the Public
    Interest
P. O. Box 298
Livingston, KY 40445
(606) 453–2315
*Information on energy
alternatives, particularly solar and
residential conservation.*

Ruth Martin
Daniel Boone Development Council
P. O. Box 431
Manchester, KY 40962
(606) 598–5127
*Technical assistance to communities
to develop energy conservation
programs.*

Robert Fehr
Co-op Extension Service Agricultural
    Engineering Dept.
University of Kentucky
Lexington, KY 40546
(606) 258–5673
*Free computer services for
energy-efficiency analysis.*

Christopher Smith
Dept. of Human Resources
Office of the Secretary
275 East Main St.
Frankfurt, KY 40621
(502) 564–7130
*Referral services for technical
assistance to low-income
organizations.*

LOUISIANA

Rose Trahan
Department of Urban and
    Community Affairs
P. O. Box 44455
Baton Rouge, LA 70804

(504) 925–3730
*Information on community action
agencies involved in energy
conservation; networking;
weatherization program.*

Frank Neelis
SMILE
501 St. John St.
Lafayette, LA 70502
(318) 234–3272
*Training, technical assistance, and
energy conservation information for
low-income persons.*

MAINE

Harvey Rosenfeld
Div. of Community Services
Statehouse Station #73
Augusta, ME 04333
(207) 289–3771
*Referral services; networking
assistance; community action agency
networking.*

Alan Lishness
Maine Audubon Society Energy
    Dept.
Gilsland Farm
118 U.S. Rt. One
Falmouth, ME 04105
(207) 781–2330
Educational materials.

Austin Bennett
Co-op Extension Service
Winslow Hall
University of Maine
Orono, ME 04469
(207) 581–2211
*Workshops; information and
assistance in weatherization, wood
heating, and appropriate technology.*

MASSACHUSETTS

Pat McGuigan
Community Development Finance
    Corp.
131 State St., Room 600
Boston, MA 02109

(617) 742–0366
*Debt and equity financing for
economic development in
conservation for eligible community
development corporations
(Massachusetts only).*

Vivian Male
Community Economic Development
  Assistance Corp.
27 School St., Room 500
Boston, MA 02108
(617) 727–0506
*Technical assistance for nonprofit
community organizations.*

MICHIGAN

Connie Williams
Community Action Agency Assoc.
451 Hollister Bldg.
Lansing, MI 48933
(517) 484–1353
*Education and training for
community conservation programs.*

MINNESOTA

Jim Solem, Exec. Dir.
Minnesota Housing Finance Agency
333 Sibley St. #200
Nalpak Bldg.
St. Paul, MN 55101
(612) 296–7615 (loan info.)
(612) 206–9807 (grant info.)
*Grants and home improvement loans
for energy conservation.*

Tom Triplett
Minnesota Project
618 East 22nd St.
Minneapolis, MN 55404
(612) 870–4700
*Technical assistance; planning
assistance; grant-writing assistance;
help in finding funding and resources;
community assistance staff for
energy.*

James Uttley
Metropolitan Council
300 Metro Square Bldg.
St. Paul, MN 55101
(612) 291–6361
*Energy-planning assistance and grant
information for local governments.*

MISSISSIPPI

Dianne Ford
Mississippi Solar Council
887 Briarwood Dr.
Jackson, MS 39211
(601) 956–4868
*Technical assistance and information
on appropriate technology;
networking.*

Brenda Morant
Gulf Coast Community Action
  Agency
P. O. Box 519
Gulfport, MS 39501
(601) 863–2233
*Referral services for information on
conservation and appropriate
technology.*

MISSOURI

Jerry Wade
Community Energy Project
College of Public and Community
  Services
Clark Hall, 7th fl.
Columbia, MO 65211
(314) 882–7503
*County extension agents provide
technical assistance on energy
conservation and appropriate
technology.*

Steve Johnson
County Energy Project
811 Cherry St. #318
Columbia, MO 65201
(314) 449–8021

Information clearinghouse for
appropriate technology and local
energy planning.

MONTANA

J. Lee Cook
Western SUN
32 South Ewing St.
Helena, MT 59601
(406) 449–4624
Technical assistance in solar systems
and design; energy management and
planning; local government assistance
with construction; management
assistance to small businesses; seminars
and workshops.

NEBRASKA

Jeff Jorgensen
Div. of Community Affairs
Dept. of Economic Development
P. O. Box 94666
301 Centennial Mall South
Lincoln, NE 68509
(402) 471–3111
Information and organizing assistance
for community energy projects;
referral services.

NEVADA

Sheldon Gordon
Nevada Assoc. of Solar Energy
   Advocates
P. O. Box 8179
University Station
Reno, NV 89507
(702) 323–1267
Publications; speakers; referral
services.

Clark Leedy
Co-op Extension Service
University of Nevada-Reno
Reno, NV 89557
(702) 784–6611

Agricultural applications of
appropriate technology and
conservation applications information.

Robert Hill
Nevada Four Corner Regional
   Commission
Capitol Complex
Carson City, NV 89710
(702) 885–4865
Funding for community energy
projects.

NEW HAMPSHIRE

Francis Gilman
New Hampshire Co-op Extension
   Service
101 Petee Hall
University of New Hampshire
Durham, NH 03824
(603) 862–1028
Technical assistance; information
materials; referral services.

Hilda Wetherbee
Total Environmental Action, Inc.
Church Hill
Harrisville, NH 03450
(603) 827–3374
Workshops; speakers; mail-order
catalog on alternative energy.

NEW JERSEY

Andrew Marshall
Bergen County CAP
8 Romanelli Avenue
South Hackensack, NJ 07606
(201) 487–3400
Assistance with CETA grant
proposals; consultation on organizing
weatherization programs; information
on domestic hot water systems,
greenhouses, and space heating.

Jamie Cromartie
Stockton Center for Environmental
   Research

Stockton State College
Pomona, NJ 08240
(609) 652–1776, ext. 211
*Energy House tours; solar and
conservation courses.*

Denise Eure
Atlantic County Energy Office
Atlantic City, NJ 08401
(609) 345–6700
*Consumer information; information
to county service organizations;
county energy auditing program;
referral services.*

NEW YORK

Brent Porter
School of Architecture
Pratt Institute
Brooklyn, NY 11205
(212) 636–3407
*Technical assistance for community
conservation projects; continuing
education courses; seminars.*

Don Price
New York Co-op Extension Service
    Energy Programs
425 Riley-Robb Hall
Cornell University
Ithaca, NY 14853
(607) 256–7733
*Energy coordinators in all counties;
technical assistance; information;
education programs.*

Kenneth Sherman
New York PIRG Citizens Alliance
295 Main St., Room 1071
Buffalo, NY 14203
(716) 847–1536
*Assistance to low-income groups in
developing weatherization programs
and appropriate-technology projects.*

NORTH CAROLINA

Paul Gallimore
Long Branch Environmental
    Education Center

Rt. 2, Box 132
Leicester, NC 28748
(704) 683–3662
*Networking; information and
technical assistance on
community-based appropriate
technology.*

NORTH DAKOTA

Bill Roath
North Dakota Energy Advocacy
    Project
2103 Lee Ave.
Bismarck, ND 58501
(701) 222–0506
*Assistance in developing low-income
community energy projects;
networking; advocating positions
favorable to low-income people as
energy consumers in matters of
public energy policy.*

OHIO

Bob Bailey & Bill Moreland
Ohio State University Energy
    Program
Oxley Hall, Room 111
1712 Neil Ave.
Columbus, OH 43210
(614) 422–5485
*Technical assistance to local
governments and businesses;
appropriate-technology center; audit
and retrofit management assistance;
workshops.*

Dick Thomas
Co-op Extension Service
2120 Fyffe Rd.
Columbus, OH 43210
(614) 422–1607
*Energy management and conservation
assistance on a county level; energy
audit programs; workshops;
literature.*

OKLAHOMA

Sherwood Washington
Dept. of Economic and Community
   Affairs
5500 North Western St.
Oklahoma City, OK 73118
(405) 840-2811
*Solar grant programs and
weatherization programs.*

Bill Zoellick
Sunspace
P. O. Box 1792
Ada, OK 74820
(405) 436-1400
*Hands-on appropriate-technology
workshops; technical assistance in
low-income appropriate-technology
applications; county energy planning.*

OREGON

Don Corson
Oregon Appropriate Technology
P. O. Box 1525
Eugene, OR 97440
(503) 683-1613
*Technical assistance to communities
in research and planning; education
programs in appropriate technology.*

PENNSYLVANIA

George Klauss
Bureau of Local Government
   Services
P. O. Box 155
Harrisburg, PA 17120
(717) 787-5177
*Training and technical assistance for
local residential groups and
government officials.*

Carolyn Boardman
Office of Community Energy
Dept. of Community Affairs
P. O. Box 156
Harrisburg, PA 17120
(717) 783-2576
*Information on community energy
education; limited funds for
community programs.*

Tom Lent
Grassroots Alliance for Solar
   Pennsylvania
3500 Lancaster Ave.
Philadelphia, PA 19104
(215) 222-0318
*Technical assistance in
community-based technologies;
networking; assistance in finding
sources of funding.*

Gary Reneker
Pennsylvania-Delaware Assoc. for
   Community Action
P. O. Box 848
Harrisburg, PA 17108
(717) 233-1075
*Referral services to locate sources of
technical assistance and assistance in
finding funding for low-income
people.*

RHODE ISLAND

Frank Lennon
Providence Corp.
204 Cranston St.
Providence, RI 02907
(401) 421-2540
*Hands-on support and technical
assistance for appropriate community
technology projects; networking;
referral services.*

Marjorie Munato
Co-op Extension Service
University of Rhode Island
Woodward Hall
Kingston, RI 02881
(401) 792-2464
*Technical assistance; workshops;
literature.*

SOUTH CAROLINA

Bill Frye
South Carolina Environmental
   Coalition (CREATE)

P. O. Box 5671
Columbia, SC 29250
(803) 799–0321
*Networking; limited technical
assistance for community groups;
referral services to locate other
sources of technical assistance.*

SOUTH DAKOTA

Karen Means
Blackhills Alliance
P. O. Box 2508
Rapid City, SD 57701
(605) 342–5127
*Outreach programs; networking;
some technical assistance; educational
services.*

Lyle Feisel
South Dakota Renewable Energy
  Assoc.
c/o South Dakota School of Mines
Rapid City, SD 57701
(605) 394–2451
*Referral services to locate sources of
technical assistance.*

Jim Ceglian/Glenn Tucker
S.T.A.T.E. Engineering Extension
South Dakota State University
Brookings, SD 57007
(605) 688–4101
*Technical information on appropriate
technology and conservation; field
visits; energy audits.*

TENNESSEE

Dr. Graham Siegel
Tennessee Valley Authority
Div. of Energy Demonstration and
  Technology
1360 Commerce Union Bank
Chattanooga, TN 37401
(615) 755–3941
*Funding proposals for innovative
conservation projects.*

Mayo Taylor
Tennessee Environmental Council

P. O. Box 1422
Nashville, TN 37202
(615) 251–1110
*Networking; general solar and
conservation information;
greenhouses.*

Brian Crutchfield
Tennessee Valley Authority
Community Development Services
110 Summer Place Bldg.
Knoxville, TN 37902
(615) 632–6980
*Energy-planning assistance in
developing community management
action plans.*

TEXAS

Jean Waugh
Dept. of Community Affairs
P. O. Box 13166
Capitol Station
Austin, TX 78711
Weatherization program; newsletter.

David Ojeda
CSA of Dimmit and LaSalle Counties
P. O. Box 488
Carrizo Springs, TX 78834
(512) 876–5219
*Assistance for low-income
community conservation projects in
the two-county area.*

Pliny Fisk
Center for Maximum Potential
  Building Systems
8604 Webberville Rd.
Austin, TX 78724
(512) 928–4786
*Technical assistance in appropriate
technology and community energy
conservation for low-income
communities.*

UTAH

Maria Tibbit/Irene Elam
Utah Community Action Assoc.
28 East 21 South #101

Salt Lake City, UT 84115
(801) 582-8181
Referral services to locate technical
assistance sources for low-income
community organizations;
help in organizing new projects.

VIRGINIA

Peter Ceperley
Solar Energy Assoc. of Northern
  Virginia
Physics Dept.
George Mason University
Fairfax, VA 22030
(703) 323-2539
Technical information on solar
applications.

Scott Forsyth
Virginia Winterization Program
P. O. Box 1997
Richmond, VA 23216
(804) 786-1798
Referral services to locate sources of
technical assistance for low-income
organizations.

WASHINGTON

J. Orville Young/Henry Waelti
Co-op Extension Service
Washington State University
AG Phase II
Pullman, WA 99164
(509) 335-2511
Technical assistance, conservation,
and appropriate-technology materials
for communities.

Birny Birnbaum
Citizens for a Solar Washington
Box 20123
Seattle, WA 98102
(206) 322-1812
Technical resources for communities
interested in developing
appropriate-technology programs.

Lucy Gorham
Neighborhood Technology Coalition

909 4th Ave.
Seattle, WA 98104
(206) 447-3625
Technical assistance; funding; project
management for community-based
appropriate-technology projects.

WEST VIRGINIA

Dr. Paul W. DeVore
Technology Education
609 Allen Hall
West Virginia University
Morgantown, WV 26506
(304) 293-3803
Statewide appropriate-technology
workshops on community
technologies and renewable resources.

Sue Sauter
West Virginia Peoples Energy
  Network
Rt. 10, Box 248C
Morgantown, WV 26505
(304) 328-5143
Newsletter; networking services.

WISCONSIN

Doug Steege
Dept. of Local Affairs and
  Development
P. O. Box 7970
Madison, WI 53703
(608) 266-6747
Grant information for housing
authorities; funding for nonprofit
organizations; technical assistance for
both profit and nonprofit groups;
home and community loan
information.

Robert Lopez
Center for Community Technology
1121 University Ave.
Madison, WI 53715
(608) 251-2207
Promotes community small-scale
appropriate technology; education;
demonstration projects; limited

technical assistance for non–Dade
County residents; library.

WYOMING

Florence Baker
Western SUN
P. O. Box 4206
University Station
Laramie, WY 82071
(800) 442–8334
(307) 766–4818
Training and technical assistance on
biomass, solar and wind projects for
community groups.

Susan Morgenstern
Dept. of Economic Planning and
   Development
Barrett Bldg.
Cheyenne, WY 82002
(307) 777–7284
Workshops and seminars; technical
assistance for community
conservation and
appropriate-technology planning.

APPENDIX EIGHT:

PUBLIC UNDERGROUND BUILDINGS YOU CAN VISIT

ALABAMA
South Central Bell Alabama Operations Center, Birmingham

CALIFORNIA
State of California Office Building, Sacramento
George Moscone Convention Center, San Francisco

DISTRICT OF COLUMBIA
Georgetown University Student Recreational Facility, Washington

MASSACHUSETTS
Pusey Library, Harvard University, Cambridge

MINNESOTA
Kelly Farm Interpretive Center, Elk River
Seward West Townhouses, Minneapolis
Walker Library, Minneapolis

Williamson Hall, Bookstore and Admissions and Records Center, University of
   Minnesota, Minneapolis
Criteria/Control Data, Terratech Center, St. Paul

MISSOURI
Monsanto Company Cafeteria, St. Louis
Visitor Center at the Jefferson National Expansion Memorial Gateway Arch,
   St. Louis

NEBRASKA
Mutual of Omaha Headquarters, Omaha

UTAH
Mormon Church Archives, Wasotch Range, Salt Lake City

VIRGINIA
Terraset Elementary School, Reston

WISCONSIN
St. Benedict's Abbey Church, Benet Lake

CANADA
Eaton Centre/Royal York Downtown Underground, Toronto, Ontario
Place Ville Marie/Bonaventure/Canada, Montreal, Quebec

APPENDIX NINE:

EARTH-SHELTERED HOME BUILDERS WHO OFTEN HAVE MODELS FOR VIEWING

CALIFORNIA
Owen W. Lindsay
P. O. Box 229
Tehachapi, CA 93561
805/822-6402

COLORADO
C. Fowler Builders
923 Squire Court
Kings View Estates
Fruita, CO 81521
303/858-9096

High Country Foundations
P. O. Box 335
Edwards, CO 81632
303/926-3230

CONNECTICUT
Richard DeNoia
31 Turner Rd.
Oakdale, CT 06370
203/447-3465

IDAHO
Sol-Earth Shelters
Robert Rickerts
6000 Sunrise Terrace
Coeur d'Alene, ID 83814
208/667-0739

ILLINOIS
Earth N' Sun Homes
David Mize—Norma Hodge

P. O. Box 107
Belknap, IL 62908
618/634-2370

American Energy
Steve Carlson
Box 164
Smithshire, IL 61478
309/325-6131

Davis Caves
William Van Hagey
200 W. Monroe St.
Chicago, IL 60606
312/346-9846

Howard Mansfield
RR #1
Palmer, IL 62556
217/526-3438

Roger Kellerman
R. #2, Box 99F
Pinckneyville, IL 62274
618/357-5446

Joan Karnuth
RR #1
Sparta, IL 62286
618/773-2592

Terry & Jana Knowles
Route #1
Bonnie, IL 62816
618/224-4001

U'BAHN Earth Homes
Gary Davis
4008 Braden
Granite City, IL 62040
618/877-4800

D & H Solartron
Delmar & Helen Perschbacher
P. O. Box 143
Okawville, IL 62271
618/243-5450

INDIANA
Indiana Solartron, Inc.
Robert E. Huey

1560 Observatory Rd.
Martinsville, IN 46151
317/831-6918

Ben H. Coffman
Box 67
Centerpoint, IN 47840
812/835-5441

Don Bitts
R. #2, Box 249
Evansville, IN 47702
812/985-3679

IOWA
A. & S. Construction
1101 N. Walnut St.
Creston, IA 50801
515/782-6406

Earth Sheltered Structures
331 N. Ellen Rd.
Cear Falls, IA 50613
319/266-2342

Terra-Home Construction
RR #1
Knoxville, IA 50138
515/842-5011; 842-6410

Underground Home Construction
Route #1
Essex, IA 51638
712/379-3624

Robert Bennett
Route #1
Lawton, IA 51030
712/944-5248

Larry Klein
Box 500
Algona, IA 50511
515/295-7373

MAINE
Sails Realty
Dale Blackie
383 U.S. Route
Scarboro, ME 04074
207/883-2590; 207/773-9980

MICHIGAN
North Country Terra-Domes
141 Grove St.
Au Train, MI 49806
906/475-9604

MINNESOTA
Underground Concrete Construction
360 Pierce Plaza Bldg.
Suite 215
N. Mankato, MN 56001
507/345-7203

Heritage International
Highway 71 & 19
Redwood Falls, MN 56283
507/637-3616

MISSOURI
Terra-Dome Corporation
14 Oak Hill Cluster
Independence, MO 64050
816/229-6000

Twenty-First Century Builders
Union Star, MO 64494
816/593-2877

Simmons & Sun Inc.
Ken Tiemann—Pam Bove
P. O. Box 1497
High Ridge Shopping Center
High Ridge, MO 63049
314/677-3969

MONTANA
Rick & Betty Weisser
Box 595
Broadus, MT 59317
406/436-2302; 406/436-2426

NEBRASKA
Steven King
Route #1, Box 102A
York, NE 68876
303/468-6389

NEW YORK
Michael Dubowyk

4623 Dewey Ave.
Rochester, NY 14612
716/865-8762

OHIO
El-De-Ohio Solartron
Richard Denton
17 W. Second St.
Fredericktown, OH 43019
614/694-9756

PENNSYLVANIA
Summerhill Estates
Craig Hallowell
RD #5
Bloomsberg, PA 17815
717/784-8997

TENNESSEE
Farmers Hardware Co.
Allen Sanders
10 Market St.
Sommerville, TN 38068
901/465-3971

TEXAS
Texas Solartron Corp.
James Powell
4105 Dorothy Rd.
Grand Prairie, TX 75052
214/641-1898

Geobuilding Systems, Inc.
Jay Swayze
P. O. Box 1556
Hereford, TX 79045
806/364-0241

Father Gregory Boensch or
    Patrick Wilson
Route 7, Box 539AA
San Antonio, TX 78221
512/626-1769

WASHINGTON
Terra-Dome Inc.
11010 18th St. NE
Lake Steven, WA 98258
206/334-1025

WISCONSIN
Encore Construction
8062 Lone Oak Court
Cross Plains, WI 53528
608/835-5604

Terrestrial Builders
Route 2, Box 191A
Alma, WI 54610
608/685-4803

W. B. Distributors
Verlyn Benoy
RR #2
Hudson, WI 54016
715/386-2213

Also, check with local architects, realtors and real estate boards for viewable earth shelters close to home.

INDEX